Early Mathematics Learning and Development

Series editor

Lyn D. English
School of STM Education, Queensland University of Technology
Brisbane, QLD, Australia

More information about this series at http://www.springer.com/series/11651

Lyn English · Tamara Moore
Editors

Early Engineering Learning

Springer

Editors
Lyn English
Queensland University of Technology
Brisbane, QLD
Australia

Tamara Moore
Purdue University
West Lafayette, IN
USA

ISSN 2213-9273 ISSN 2213-9281 (electronic)
Early Mathematics Learning and Development
ISBN 978-981-10-8620-5 ISBN 978-981-10-8621-2 (eBook)
https://doi.org/10.1007/978-981-10-8621-2

Library of Congress Control Number: 2018934915

Printed on acid-free paper

This Springer imprint is published by the registered company Springer Nature Singapore Pte Ltd. part of Springer Nature
The registered company address is: 152 Beach Road, #21-01/04 Gateway East, Singapore 189721, Singapore

Preface

Early Engineering Learning is a volume within the Springer series,*Early Mathematics Learning and Development.* The collection of volumes published in this series explores a range of perspectives on young children's developments in mathematics and allied fields. One such field that draws on and fosters young children's mathematical capabilities is engineering. Despite early childhood being a period of experimentation and curiosity with the natural world and its myriad challenges, children's natural propensity for engaging in engineering experiences remains untapped.

As the chapters in this volume illustrate, we need to capitalize on children's skills as independent problem solvers who relish challenges, persevere in the face of failure, and learn both from what "works" and what does not. Educators, including parents, need to be cognizant of how children's talents can be harnessed and enriched to sow the seeds of engineering education.

Engineering has received almost no attention in the pre-K and beginning school years, even though the need for quality STEM education across all age levels is advocated by many nations. The "E" in STEM tends to be ignored in these significant formative years when an interest in and awareness of engineering and engineering design processes can be fostered. The early years of a child's life are too valuable to deprive them of the rich learning opportunities that engineering can offer.

Because engineering shapes so much of our actual and virtual worlds, it is an ideal discipline to both link and promote the varied capabilities young children bring to informal and formal learning environments. The chapters in this volume attest to the rich opportunities engineering affords. The authors report on research illustrating several intervention programs, together with assessment frameworks, which aim to facilitate beginning engineering learning. These include the use of robotics as a playful vehicle for fostering engineering, computer science, and mathematics (Chap.11), and the incorporation of literature as a familiar and meaningful basis for learning across the entire STEM curriculum (e.g., Chaps.9 and 10). A focus on spatial skills including intervention experiences, which are so important to success in engineering, is also featured (Chap. 5). Other chapters highlight the nature and role of engineering design processes and habits of mind, which are not unique to the engineering field, rather, are applicable across the curriculum.

Engineering design has been described as the "disciplinary glue" (Chap. 9) that assists children to apply their learning in STEM to an engineering design challenge. Indeed, the practice of engineering inherently requires the practitioner to call upon other disciplinary knowledge in order to solve engineering problems. Engineering design challenges are usually described as strongly iterative, open to many possible solutions, and engendering thinking processes or "habits of mind." These thinking skills underline design processes and include systems thinking, innovative problem finding and solving, visualizing, and collaborating and communicating (as addressed in Part I and Chap. 13).

Readers might notice that this volume comprises only 13 chapters, a reflection of the embryonic nature of the field. As such, the volume presents a seminal set of research-based studies that provide empirical evidence of what can be achieved in implementing engineering experiences in early childhood. *Early Engineering Learning* raises the profile of an overlooked discipline that is as natural to early childhood as is mathematics, science, and technology. A STEM agenda is not complete without engineering, nor is a child's early learning and development. We can no longer ignore this core discipline.

Brisbane, Australia Lyn English

Contents

Chapter 1
Early Engineering: An Introduction to Young Children's Potential

Lyn D. English

Abstract *Early Engineering Learning* comprises two main sections presenting a mix of research studies, theoretical advancements, and classroom empirical examples. As such, this book provides a rich resource for researchers, policy makers, curriculum developers, and classroom teachers alike. Engineering learning is a significant yet underrepresented field in early education, despite being one of the most practical and real-world domains that all children can engage in. As evident in the chapters of this book, young children are eminently capable of solving engineering-based problems; indeed, they do this on a daily basis. Engineering education integrates readily and meaningfully not only within the other STEM domains, but also with literature and the arts more broadly. Various approaches to early engineering learning are showcased throughout this book, with engineering design processes and habits of mind featured prominently. Not only are these design and thinking processes foundational to early engineering but can also enhance learning across several other disciplines.

The chapters of *Early Engineering Learning* comprise a mix of research studies, theoretical advancements, and empirical examples for classroom use. As such, this book provides a rich resource for researchers, policy makers, curriculum developers, and classroom teachers alike. Engineering learning is a significant yet underrepresented field in early education, despite being one of the most practical and real-world domains that students of all ages can experience with success and enjoyment. Indeed, young learners are natural engineers, as has often been stated (e.g., https://www.eie.org/eie-curriculum/why-engineering-children).

Despite the ubiquity of engineering throughout our environment, education has yet to capitalize fully on the domain's potential for early learning—in essence, we are ignoring young children's capabilities for engaging in engineering experiences.

The original version of this chapter was revised: XX entries are removed. The correction to this chapter is available at https://doi.org/10.1007/978-981-10-8621-2_14

L. D. English (✉)
Faculty of Education, Queensland University of Technology, Brisbane, QLD , Australia
e-mail: l.english@qut.edu.au

L. English and T. Moore (eds.), *Early Engineering Learning*, Early Mathematics Learning and Development, https://doi.org/10.1007/978-981-10-8621-2_1

As evident in the chapters of this book, young children are eminently capable of solving engineering-based problems; indeed, they do this on a daily basis. Young children's natural curiosity, inquiry, and desire to explore their world form not only the cornerstone of early childhood development (Brophy, Klein, Portsmore, & Rogers, 2008), but also "a key component of thinking like an engineer" (Elkin, Sullivan, & Bers, Chap. 11; Tippett & Milford, 2017). One only has to observe how young children investigate, experiment, manipulate, and create with everyday objects to realize how they are engaging in the foundations of engineering education. As children are exposed to these foundations, they are developing core discipline knowledge that enriches not only their mathematics, science, and technology curricula but also other content areas. As Petroski (2016) highlighted, "Engineering is not an end in itself. It operates in a moral, social, economic, and aesthetic context" (p. 21).

Unfortunately, we have neglected the "E" in our STEM education for too long (Di Francesca, Lee, & McIntyre, 2014; English & King, 2017; Moore et al., 2014). Despite a lack of specifically developed resources and associated teacher professional development, as I note in the final chapter, the contributions of engineering to our world need greater recognition—and what better way to start than through nurturing an early awareness of how engineers and engineering shape our world.

The present chapters are arranged in two main sections, in addition to introductory and concluding chapters. In addressing early engineering learning from Pre-K through to the early years of formal education, the chapters in the first section focus primarily on engineering thinking, design, and habits of mind, while those in the second section target early engineering curriculum development. There is naturally some overlap in these sections as curriculum and resource development necessarily takes into account engineering design and thinking.

1.1 Why Focus on Engineering Thinking, Design, and Habits of Mind?

The chapters in the first section lay frameworks for early engineering learning, with studies ranging from capitalizing on spontaneous play as opportunities for introducing engineering (e.g., Chaps. 4 and 6) through to fostering early spatial skills as a core habit of mind in both engineering and STEM more broadly (Chap. 5). Combined with chapters presenting examples of observation protocols and specifically designed assessment tools (e.g., Chaps. 6 and 7), the first section draws together a range of research and classroom tested ideas that collectively pave the way for further studies and advancement.

As one peruses the chapters in each section, it will become apparent that engineering design processes feature prominently. Some might even argue that there is too much emphasis on these processes, yet as Tank, Moore, Gajdzik, and Sanger (Chap. 9) aptly state that engineering design is the "interdisciplinary glue" in STEM education. Research has indicated how engineering design, a core construct in the discipline, enables learners to appreciate that there are multiple ideas and approaches to solving complex problems with more than one solution possible, that numerous tools and representations can be used in different ways to produce a desired end-product, and that it is acceptable for initial designs to "fail" necessitating redesign and improvement (e.g., Dorie, Cardella, & Svarovsky, 2014; English & King, 2017; Tank et al., Chap. 9). Indeed, repeated studies have illustrated how engineering design processes provide a meaningful tool for all learners, across ages and grade levels, in solving not only engineering-based problems but also numerous other real-world challenges. The chapters in Sect. 1.1 provide many examples that collectively convey the message that young children have substantial potential for engaging in engineering thinking, applying engineering design processes, and displaying foundational habits of mind. Although providing somewhat similar evidence of these capabilities, the chapters nevertheless reinforce the urgent need to attend to *all* of STEM in early education, not just mathematics and science.

Apart from advancing engineering learning, these first section chapters contribute to early curriculum development more broadly—not only with respect to STEM education but also the arts including literature, as addressed in the second section. The broadening of STEM to STEAM education is gaining in popularity as the advantages of incorporating components of the arts are recognized (as revisited in Chap. 13).

1.2 Early Engineering Curriculum Development

As a seminal, early engineering program, *Engineering is Elementary* (Chap. 8; Cunningham & Hester, 2007) builds specifically on designed engineering stories and has laid the groundwork for many subsequent engineering programs, as indicated in Chap. 8 (Cunningham, Lachapelle, & Davis). The many contributions of engineering to our environment and our lives more broadly are reflected in the ease with which the discipline can be integrated within early educational programs. Engineering shares more than mathematics and science components—it lends itself to a range of literature and to the natural problem-solving situations that occur in our everyday lives. To cite Petroski (2016) again, "Engineers have come to be recognized as the creative people who bring us innovations like the smartphone, the personal computer, the internet, and the world wide web" (p. 21). The world revolves around these technological innovations, but do we stop to think of those creative engineers responsible for their development?

As young children interact with these technological tools, they too can begin to appreciate the powerful ways that engineers enhance our world. Indeed, our future team of engineering students need to be nurtured from a young age, at a time when children's curiosity is at its peak. Sparking such interest can ideally begin with support from literature, as studies in the second section demonstrate. Portsmore and Milto (Chap. 10), for example, discuss their *Novel Engineering* program, which replaces real-world clients and contexts with those from popular literary texts as a basis for creating engineering design challenges. Children draw information from the given literary text in identifying engineering problems, where story characters are considered as clients and details from the story are used to impose constraints, as solutions are developed for the characters' problems. The numerous other examples illustrate the need to utilize more the power of literature, an often ignored resource that can enrich so many disciplines (Luedtke & Sorvang, 2017).

Along with engineering, technology learning in the younger years requires further research and curriculum development. While there is an increasing focus on early technology especially coding (e.g., Fessakis, Gouli, & Mavroudi, 2013), the links with engineering have been underrepresented. In Chap. 11, Elkin, Sullivan, and Bers provide innovative approaches to developing foundational engineering and computer science concepts. They present insightful anecdotes illustrating how robotics can serve as a playful medium to develop these concepts. Educators with little to no prior engineering experiences were able to successfully integrate robotics with traditional early childhood curriculum content such as literacy and science. Their vignettes highlight the different approaches teachers took in introducing robotics within their classrooms and how they utilized the engineering design process as a teaching tool applicable to many subject areas, not just STEM. With the increasing availability of robotic kits for young children, such as KIBO described in Chap. 11 (www.kinderlabrobotics.com), numerous opportunities now exist for early educators to explore the learning potential of robotics. With such potential extending beyond just STEM to other content domains, robotics can readily enrich existing curricula, as Elkin et al. explore. Furthermore, early technology experiences can act as catalysts for social and emotional skill development across a diverse range of students, as seen in Elkin et al.'s study for whom English was not the first language for over a third of their participants.

Early engineering education research and development have a considerable distance to go. The chapters in this book present several avenues for traveling this distance, but obstacles need overcoming. Two of the several challenges we face with respect to research, policy, and curriculum development include increasing awareness of young children's competencies in early engineering, and enhancing teacher resources and professional development opportunities, as noted in the last chapter. With respect to the latter, curriculum resources need to be integrated within the regular curriculum, otherwise engineering education will likely be viewed as another "add-on" to be squeezed into an already tight curriculum. Fortunately, engineering lends itself easily to such integration, linking not only with the remaining STEM disciplines but also with other domains especially literature. It is to be hoped that curriculum developers across the disciplines can capitalize on the many contributions of early engineering education. The chapters in this book provide rich starting points.

References

Brophy, S., Klein, S., Portsmore, M., & Rogers, C. (2008). Advancing engineering education in P-12 classrooms. *Journal of Engineering Education, 97*(3), 369–387.

Cunningham, C. M., & Hester, K. (2007). *Engineering is elementary: An engineering and technology curriculum for children*. In ASEE Annual Conference & Exposition, Honolulu, HI.

Di Francesca, D., Lee, C., & McIntyre, E. (2014). Where is the "E" in STEM for young children? *Issues in Teacher Education, 23*(1), 49–64.

Dorie, B. L., Cardella, M. E. & Svarovsky, G. N. (2014). *Capturing the design thinking of young children interacting with a parent*. Paper presented at the 121st SEE Annual Conference & Exposition, Indianapolis.

English, L. D., & King, D. T. (2017). Engineering education with fourth-grade students: Introducing design-based problem solving. *International Journal of Engineering Education, 33*(1B), 346–360.

Fessakis, G., Gouli, E., & Mavroudi, E. (2013). Problem solving by 5–6 years old kindergarten children in a computer programming environment: A case study. *Computers & Education, 63*, 87–97.

Luedtke, M., & Survang, K. (2017). Using children's literature to enhance math instruction in K-8 classrooms. In L. Jao & N. Radakovic (Eds.), *Transdisciplinarity in mathematics education: Blurring disciplinary boundaries* (pp. 47–71). Berlin: Springer.

Moore, T. J., Stohlmann, M. S., Wang, H., Tank, K. M., Glancy, A. W., & Roehrig, G. H. (2014). Implementation and integration of engineering in K-12 STEM education. In S. Purzer, J. Strobel, & M. Cardella (Eds.), *Engineering in Pre-College Settings: Research into Practice* (pp. 35–60). West Lafayette, IN: Purdue University Press.

Petroski, H. (2016). Refractions: Feeling superior? *PRISM*, September, 21.

Tippett, C. D., & Milford, T. M. (2017). Findings from a pre-kindergarten classroom: Making the case for STEM in early childhood education. *International Journal of Science and Mathematics Education*, (Suppl 1), 67–86.

References

[illegible]

Part I
Engineering Thinking and Habits of Mind

Chapter 2
Engineering in the Early Grades: Harnessing Children's Natural Ways of Thinking

Tamara J. Moore, Kristina M. Tank and Lyn English

Abstract This chapter explores engineering as it applies to students in the early grades. First, we consider engineering as a STEM foundation. We then address ways in which we can provide supportive learning environments for early engineering learning. As part of such environments, we examine how we can build on intrinsically interesting problems. In exploring ways of harnessing young learners' natural ways of thinking, we consider the role of play in early engineering learning and how we can capitalize on this play. The integration of engineering within the early curriculum is then reviewed, followed by a summary of perspectives on ways in which engineering is developmentally appropriate for, and beneficial to, young learners.

Engineering is a multifaceted field that draws not only from related disciplinary domains such as mathematics and science, but also from disciplines that serve to make engineering solutions more practical or desirable such as economics, social studies, and the arts. Technological developments such as the iPhone, robotics, and 3-D printing, all involve major engineering inputs. Young children are very much a part of our engineered world, interacting daily with the products of engineering and technology.

On entering kindergarten, children already have sophisticated ways of thinking about the world based largely on their own experiences (Baillargeon, 1994; Cohen & Chashon, 2006), which serve as a springboard for their future learning and development (Inagaki & Hatano, 2006; NRC, 2012). Early engineering education

T. J. Moore (✉)
Purdue University, West Lafayette, IN, USA
e-mail: tamara@purdue.edu

K. M. Tank
Iowa State University, Ames, IA, USA
e-mail: kmtank@iastate.edu

L. English
Queensland University of Technology, Brisbane, QLD, Australia

L. English and T. Moore (eds.), *Early Engineering Learning*, Early Mathematics Learning and Development, https://doi.org/10.1007/978-981-10-8621-2_2

falls naturally within such experiences. With its focus on iterative thinking—that is, trying something, testing it, learning from what does not go well, and trying again—as well as working in teams and communicating current ways of thinking, engineering is an ideal avenue for enriching and extending young children's natural talents. As Lippard, Lamm, and Riley (2017) noted, "... Pre-kindergarten children are primed for engineering thinking" (p. 455). Furthermore, engineering provides a platform for young children to be introduced to technology, not just as digital media, but as all aspects of the designed world. Children inherently alter their environment to fit their needs. These alterations are the beginnings of engineering thinking, which can promote structured decision making within a specified engineering context.

This chapter explores engineering as it applies to students in the early grades. First, we consider engineering as a STEM foundation. We then address ways in which we can provide supportive learning environments for early engineering learning. As part of such environments, we examine how we can build on intrinsically interesting problems. As we continue to explore ways of harnessing young learners' natural ways of thinking, we consider the role of play in early engineering learning and how we can capitalize on this play.

2.1 Engineering as a STEM Foundation

Research over many years has revealed that young children have sophisticated minds and a natural eagerness to engage in a range of mathematical and scientific activities a good deal earlier than previously thought (Perry & Dockett, 2013; English & Mulligan, 2013; English, 2013; Lehrer & English, 2018). Children enter kindergarten with surprising ways of thinking about the world they experience, which can be used to promote problem-solving and build understanding in the early grades (Baillargeon, 1994; Cohen & Chashon, 2006). Indeed, a range of studies in prior-to-school and early school settings have revealed how young learners possess cognitive abilities which, with appropriately designed and implemented learning experiences, can enable forms of reasoning not typically seen in the early years (e.g., Clements, Sarama, Spitler, Lange, & Wolfe, 2011; English, 2012; Inagaki & Hatano, 2006; Lehrer & English, 2018; Lehrer & Schauble, 2015; Moss, Bruce, & Bobis, 2016; Perry & Dockett, 2008). For example, young children can abstract and generalize mathematical and scientific ideas much earlier, and in more complex ways, than previously considered. These sophisticated ways of thinking and reasoning in young children provide a foundation that can be used to not only facilitate early engineering knowledge and skills but also to support early learning across other content areas, such as mathematics and science.

2.2 Providing Supportive Learning Environments

In efforts to provide supportive and facilitating environments for young learners, educators frequently overlook the potential contributions of engineering. The discipline lends itself effectively to nurturing young children's natural ways of thinking, while at the same time promoting engineering knowledge, thinking skills, and productive problem-solving. When looking at how young learners explore, interact, and think about their world, it is important to consider different ways of shaping environments that facilitate this growth. The importance of providing learning environments that capitalize on young children's natural capabilities was emphasized in Moss, Bruce, and Bobis' (2016) review of challenges and developments in early mathematics learning. Their review indicated how the development and implementation of enriched and expanded programs in the early years are being increasingly recognised as crucial for future achievement, with associations such as the National Council of Teachers of Mathematics and the National Association for the Education of Young Children (NAEYC/NCTM, 2009) strongly endorsing such programs.

Research on early science learning has also revealed young children's innate talents in the STEM fields. Studies have highlighted their fundamental understanding of observational phenomena and knowledge about the natural world that results from investigating and exploring their environment (Eshach & Fried, 2005; French, 2004). This innate curiosity and sense of wonder about the world around them leads to a natural tendency to observe, explore, and try to explain their everyday experiences (Eshach & Fried, 2005). Even before entering school, young children are able to recognize patterns and then use those causal and relational patterns to reason about living things and natural phenomena (Inagaki & Hatano, 2006). As these young learners are exploring their environment and acquiring knowledge, it is their personal experiences that form the foundation for their understanding of and interactions with the natural and manufactured world (French, 2004).

While the focus for these experiences is often the natural environment, there are also interactions within the designed, human-crafted world. Engineering comes to the fore here, with opportunities to build upon and engage children's desire to make things and to learn how various objects work (Brophy, Klein, Portsmore, & Rodgers, 2008). Engineering also provides an avenue for young children to be introduced to technology, not just as digital media, but also as all parts of the designed world.

In sum, with children's curiosity and motivation to explore their world, they inherently alter their environment to fit their needs. These alterations are the beginnings of engineering thinking. In harnessing children's innate ways of thinking within their environments, engineering experiences can foster structured decision making within a specified engineering context.

2.3 Building upon Intrinsically Interesting Problems

Along with this natural curiosity to learn about and investigate their world, young children spend substantial time troubleshooting and designing as they explore various problems in their surroundings. For example, Bairaktarova, Evangelou, and Brophy (2011) observed students during exploratory play where they intentionally modified existing structures or artifacts to solve a problem. They also continually tested the limits of their experimental designs by adding "one more block" and observing what works and what does not (Bairaktarova, Evangelou, & Brophy, 2011). These examples illustrate the idea that young children often spend time-solving problems, which tend to be open-ended or ill-structured, with parallels to problems and problem-solving skills that are characteristic of engineering (Brophy et al., 2008; Watkins, Spencer, & Hammer, 2014). In fact, many common behaviors expressed by young children, such as their desire to ask questions, explore, and develop creative solutions, resemble highly desirable traits within engineering. As such, these behaviors can be viewed as precursors to engineering and engineering thinking (Brophy & Evangelou 2007; Lippard, Lamm, & Riley, 2017; Van Meeteren & Zan, 2010).

Engineering experiences can also help students and teachers move beyond simply solving problems to emphasizing a level of intentionality and motivation in their actions. Such intentionality was revealed in Fleer's (2000) study as preschool children were able to plan, design, and then use their prior experience with materials to predict which materials they need for their designs. Additionally, young children have been shown to communicate their plans for constructing products with some level of intention, which has been shown to extend even to the evaluation of their designs (Johnsey, 1995; Brophy & Evangelou, 2007; Bagiati, 2011). Furthermore, studies of pre-schoolers engaging in block-building and other free-play activities have identified instances where students solved problems and pursued goals that met a certain set of constraints and engaged in iterative cycles of problem-solving in achieving the goal (Bairaktarova et al., 2011; Brophy et al., 2008). In essence, early engineering experiences build upon young children's inherent desire to solve problems and alter their environment to fit their needs, while also promoting early problem-solving skills and encouraging progress beyond just solving problems.

2.4 Providing a Vehicle for Curriculum-Based Child-Centered Play

The important role of play in early education has a long history (e.g., see Moss et al., 2016). It is well recognized that play can foster the development of positive dispositions and habits of mind including curiosity, creativity, diverse problem-solving, and communicating ideas and emotions (e.g., Ginsburg, 2009; NAEYC, 2010). On the other hand, there are debates regarding the extent to which specific disciplines such as mathematics and science should be learned through a play-based approach.

For example, a common belief has been that all mathematics learning should emerge from child-directed play. Although the importance of young children being actively involved in constructing their mathematics and science knowledge cannot be disputed, there remain questions about the appropriate learning environments and supports needed to maximize such learning. As Moss et al. (2016) indicated, there are potential limitations in relying on unguided play for deep early learning in mathematics. While it is well acknowledged that play has an important role in young children's discipline learning (e.g., Perry & Dockett, 2008; Sarama & Clements, 2009), such an approach does not ensure maximum mathematical development (de Vries, Thomas & Warren, 2010, p. 717). Sarama and Clements (2009) further illustrated how desired mathematical concepts are unlikely to be developed when children play with mathematics-related materials and objects solely by themselves.

Referring back to our discussion on environments that facilitate early engineering learning, it is worth considering briefly Moss et al.'s (2016) review of new developments in the field. Citing a "playful pedagogy" approach, Moss et al. report on research suggest that a "middle ground" between free play and direct instruction may be most effective in improving access to a more in-depth and broader array of early mathematics learning opportunities. Such an approach integrates a child-centered play mode with curricular goals and allows children to control their learning to a large degree. Baroody's (2006) early years continuum of pedagogies for mathematics features four main aspects, ranging from traditional direct instruction, to guided discovery learning through an adult-initiated task, through to flexible guided discovery learning by means of a child-initiated task, and finally, unguided discovery involving a child-initiated task. Not surprisingly, Baroody's classroom observations revealed that the mid-way approaches, namely discovery and flexible guided discovery, were the most promising for fostering mathematics learning, although each the four approaches had an important role in early childhood environments. Additionally, Moomaw's (2014) findings illustrate the rich and varied science, math, and engineering experiences that young children routinely encounter in a high-quality early childhood classroom, where learning and play are intentionally combined. Consequently, when examining early childhood experiences within the frame of a play-based approach, early engineering provides an opportunity to establish environments that support intentional and explicit connections between science and mathematics content and the free-play environments that are frequently seen and promoted in early childhood classrooms (Bairaktarova et al., 2011).

2.5 Integrating Engineering Within the Regular Curriculum

Engineering in the early grades should include varied opportunities for students to experience examples of engineering and engage in engineering design and thinking activities that allow them to begin to understand engineering as a broad discipline.

At the same time, such activities should provide ways for them to dig deeply into aspects of the domain. Developing engineering thinking is not a straightforward task, nor is incorporating the domain within an already overcrowded curriculum (Lippard et al., 2017).

One approach to addressing these difficulties is through integrating engineering within the other curriculum disciplines, as illustrated in several chapters in this book. Indeed, engineering is seen as providing a foundational, cross-disciplinary link that contextualizes students' mathematics, science, and technology learning (e.g., English, 2017; Moore et al., 2014; Zawojewski, Diefes-Dux, & Bowman, 2008). Although engineering design processes provide important foundational links across the STEM disciplines and enable students to appreciate how multiple ideas, approaches, and tools can be applied to complex problems involving more than one solution (Purzer, Hathaway Goldstein, Adams, Xie, & Nourian, 2015), their multiple applications in the curriculum are not being acknowledged adequately.

Although frameworks for integrating engineering learning within the early school years are not prolific, Bryan, Moore, Johnson, and Roehrig's (2016) "STEM Roadmap" provides a rich source of ideas. Within their framework, engineering design and practices form a key component in linking science and mathematics, with five core instructional features advanced: (1) the content and practices of one or more of the science and mathematics disciplines comprise some of the primary learning goals; (2) engineering practices and engineering design of technologies, either as the context or the intentional learning content or both, serve to integrate the disciplines; (3) the scientific and mathematical concepts that are required for the engineering components include design justification; (4) the development of twenty-first-century skills is highlighted; and (5) the instructional context requires solving a real-world problem or task through collaborative groups.

Importantly, as emphasized by both Bryan et al. (2016) and Lippard et al. (2017), STEM integration needs to be "intentional" and "specific" with consideration given to both content and context. Three forms of STEM integration are: (a) content integration where learning experiences have multiple STEM learning objectives, (b) integration of supporting content where one area is addressed (e.g., mathematics) in support of the learning objectives of the main content (e.g., science), and (c) context integration where the context from one discipline is used for the learning objectives from another (Moore & Hughes, 2019; Bryan et al., 2016). Although the integration of supporting content is frequent, it appears not to be applied in a way that effectively extends this content (Bryan et al., 2016). Unfortunately, the broad contributions of engineering education to early children's learning are not being adequately recognized in many nations. Yet as the report *Engineering in K-12 Education* (National Academy of Engineering and National Research Council, 2009) stressed, "In the real world, engineering is not performed in isolation—it inevitably involves science, technology, and mathematics. The question is why these subjects should be isolated in schools" (pp. 164–165).

2.6 Perspectives on Early Engineering

This chapter has reviewed several different ways in which engineering is developmentally appropriate for, and beneficial to, early engineering learners. Throughout the remainder of this book, there are chapters that provide different perspectives on engineering in the early grades.

Engineering thinking is broader than engineering design alone. While engineering thinking includes engineering design, it also highlights that engineers are independent thinkers who seek out new knowledge when solving problems (Moore et al., 2014). Often the ways in which engineers think beyond just design are called engineering habits of mind. According to the Royal Academy of Engineering in the UK (Lucas, Hanson, Bianchi, & Chippindall, 2017), the core engineering thinking attribute is "making 'things' that work and making 'things' work better" (p. 5). The report further separates this core attribute into the engineering habits of mind that include improving, systems thinking, adapting, problem-finding, creative problem-solving, and visualizing. Other definitions of engineering habits of mind also include optimism, collaboration, communication, and ethical considerations (National Academy of Engineering & National Research Council, 2009). It has been argued that when engineering design, combined with engineering thinking, is made an explicit outcome of learning, students have increased opportunities to become independent and reflective thinkers with the skills needed to integrate multiple ideas in solving problems (Bryan et al., 2016; Lucas et al., 2014). Furthermore, engineering design processes serve to help students make connections between engineering and the other STEM disciplines as well as assist students to recognize that engineers think through problems in a systematic way.

As we consider how to implement engineering with young learners and different ways to harness students' natural ways of thinking, we see there are several different perspectives on introducing students to engineering. From the work presented in this book, we see that open-ended challenges are helpful in fostering problem-finding and creative problem-solving, that play in engineering can foster all of the engineering habits of mind, and that more formal design is effective in nurturing specific learning objectives particularly with content outside of engineering. The chapters demonstrate that there should not just be one approach to engineering learning in the early grades, but rather a mixture of learning opportunities that should provide a more well-rounded education.

2.7 Concluding Points

With the inclusion of engineering in the early grades, it is important to examine how engineering, engineering design, and engineering thinking can facilitate student learning and support teaching and learning across content areas. When presented in developmentally appropriate ways, early engineering can help young learners by supporting the development of natural ways of thinking into productive problem-solving. Young children come to the classroom with ideas about the natural and designed world that are developed as they explore, test, and modify the world around them. Early engineering provides an environment that encourages this curiosity and motivation to explore and alter the world around them. As students are investigating their world, they often engage in creative problem-solving as they try to better understand why things work and how to improve them. Incorporating engineering into early childhood classrooms builds on these problem-solving opportunities that are intrinsically interesting for young children and provide a structure that can help them move beyond simply solving problem to a level of intentionality with their problem-solving. Engineering also has a focus on iterative thinking that encourages children to engage in multiple rounds of investigating, testing, and modifying these problems that encourage deeper understanding. Many of these problem-solving opportunities occur as students are engaged in child-centered learning and play. Early engineering provides a vehicle for facilitating more intentional content connections and content learning within this child-centered approach common in the early grades. Therefore, when thinking about the bigger picture of student learning and development in the younger grades, early engineering experiences can build up contexts that are realistic and motivating for young children while also provide a way to integrate learning across subjects.

While it is important to recognize the affordances that early engineering can provide, it is also important to note the multiple ways in which engineering and engineering design can be presented in a developmentally appropriate manner for young learners. There are several approaches to engineering that can be used with young children; these multiple perspectives allow students to think about and engage with engineering, engineering design, and engineering thinking in different ways. These different ideas around engineering should not be in competition with one other, but should be more of a multiple representations approach to engineering for young children, affording them the opportunity to engage more deeply with engineering content, skills, and habits of mind.

References

Bagiati, A. (2011). *Early engineering: A developmentally appropriate curriculum for young children*. Doctoral dissertation, Purdue University.

Baillargeon, R. (1994). How do infants learn about the physical world? *Current Directions in Psychological Science, 3*(5), 133–140. https://doi.org/10.1111/1467-8721.ep10770614.

Bairaktarova, D., Evangelou, D., & Brophy, S. (2011). Early engineering in young children's exploratory play with tangible materials. *Children, Youth and Environments, 21*(2), 212–235.

Baroody, A. J. (2006). Mastering the basic number combinations. *Teaching children mathematics, 23,* 22–31.

Brophy, S. P., & Evangelou, D. (2007). Precursors to engineering thinking. In *Proceedings of the Annual Conference of the American Society of Engineering Education*, Honolulu, HI.

Brophy, S., Klein, S., Portsmore, M., & Rodgers, C. (2008). Advancing engineering education in P-12 classrooms. *Journal of Engineering Education, 97*(3), 369–387.

Bryan, L. A., Moore, T. J., Johnson, C. C., & Roehrig, G. H. (2016). Integrated STEM education. In C. C. Johnson, E. E. Peters-Burton, & T. J. Moore (Eds.), *STEM road map: A framework for integrated STEM education* (pp. 23–37). New York, NY: Routledge.

Clements, D. H., Sarama, J., Spitler, M. E., Lange, A. A., & Wolfe, C. B. (2011). Mathematics learned by young children in an intervention based on learning trajectories: A large-scale cluster randomized trial. *Journal for Research in Mathematics Education, 42*(2), 127–166.

Cohen, L. B., & Chashon, C. H. (2006). Infant cognition. In W. Damon & R. M. Lerner (Eds.), *Handbook of Child Psychology, Set, 6th Edition* (Chap. 5, Vol. 2). Hoboken, NJ: Wiley.

de Vries, E., Thomas, L., & Warren, E. (2010). Teaching mathematics and play based learning in an Indigenous early childhood setting: Early childhood teachers' perspectives. In L. Sparrow, B. Kissane, & C. Hurst (Eds.), *Shaping the future of mathematics education—Proceedings of the 33rd annual conference of the Mathematics Education Research Group of Australasia* (pp. 719–722). Fremantle, WA: MERGA.

English, L. D. (2012). Data modelling with first-grade students. *Educational Studies in Mathematics Education, 81*(1), 15–30.

English, L. D. (2013). Reconceptualising statistical learning in the early years. In L. D. English & J. Mulligan (Eds.), *Reconceptualising early mathematics learning* (pp. 67–82). Dordrecht: Springer.

English, L. D. (2017). Advancing elementary and middle school STEM education. *International Journal of Science and Mathematics Education, 15*(1), 5–24. https://doi.org/10.1007/s10763-017-9802-x.

English, L. D., & Mulligan, J. (2013). Perspectives on reconceptualising early mathematics learning. In L. D. English & J. Mulligan (Eds.), *reconceptualising early mathematics learning* (pp. 1–4). Dordrecht: Springer.

Eshach, H., & Fried, M. N. (2005). Should science be taught in early childhood? *Journal of Science Education and Technology, 14*(3), 315–336. https://doi.org/10.1007/s10956-005-7198-9.

Fleer, M. (2000). Working technologically: Investigations into how young children design and make during technology education. *International Journal of Technology and Design Education, 10*(1), 43–59. https://doi.org/10.1023/a:1008923410441.

French, L. (2004). Science as the center of a coherent, integrated early childhood curriculum. *Early Childhood Research Quarterly, 19*(1), 138–149. https://doi.org/10.1016/j.ecresq.2004.01.004.

Ginsburg, H. P. (2009). The challenge of formative assessment in mathematics education: Children's minds, teachers' minds. *Human Development, 52*(2), 109–128. https://doi.org/10.1159/000202729.

Inagaki, K., & Hatano, G. (2006). Young children's conception of the biological world. *Current Directions in Psychological Science, 15*(4), 177. https://doi.org/10.1111/j.1467-8721.2006.00431.x.

Johnsey, R. (1995). The place of the process skill making in design and technology: Lessons from research into the way primary children design and make. In *Proceedings of the IDATER95: International Conference on Design and Technology Educational Research and Curriculum Development*. Loughborough, UK: Loughborough University of Technology.

Lehrer, R. & English, D. (2018). Introducing children to modeling variability. In D. Ben-Zvi, J. Garfield., & K. Makar (Eds.), *International handbook of research in statistics education* (pp. 229-260). Dordrecht, the Netherlands: Springer.

Lehrer, R., & Schauble, L. (2015). Developing scientific thinking. In L. S. Liben & U. Müller (Eds.), *Cognitive processes*: *Vol. 2. Handbook of child psychology and developmental science* (7th ed.). Hoboken, NJ: Wiley (Editor-in-chief: R. M. Lerner).

Lippard, C. N., Lamm, M. H., & Riley, K. M. (2017). Engineering thinking in pre-kindergarten children: A systematic literature review. *Journal of Engineering Education, 106*(3), 454–474.

Lucas, B., Claxton, G., & Hanson, J. (2014). *Thinking like an engineer: Implications for the education system*. London, UK: Royal Academy of Engineers. Retrieved from http://www.raeng.org uk/policy/education policy/learning-and-teaching.

Lucas, B., Hanson, J., Bianchi, L., & Chippindall, J. (2017). *Learning to be an engineer: Implications for the education system.* London, UK: The Royal Academy of Engineering. Retrieved from http://www.raeng.org.uk/policy/education-policy/learning-and-teaching.

Moomaw, S. (2014). Don't forget to play. *School Science & Mathematics, 114,* 51–52. https://doi.org/10.1111/ssm.12056.

Moore, T. J., & Hughes, J. E. (2019). Chapter 12: Teaching and learning with technology in science, engineering, and mathematics. In M. D. Roblyer, & J. E. Hughes (Eds.), *Integrating educational technology into teaching* (8th ed.). New York, NY: Pearson.

Moore, T. J., Glancy, A. W., Tank, K. M., Kersten, J. A., Smith, K. A., & Stohlmann, M. S. (2014). A framework for quality K-12 engineering education: Research and development. *Journal of Pre-College Engineering Education Research, 4*(1), 1–13. https://doi.org/10.7771/2157-9288.1069.

Moss, J., Bruce, C. D., & Bobis, J. (2016). Young children's access to powerful mathematics ideas: A review of current challenges and new developments in the early years. In L. D. English & D. Kirshner (Eds.), *Handbook of international research in mathematics education* (Third ed., pp. 153–190). New York: Routledge.

National Academy of Engineering, & National Research Council. (2009). *Engineering in K-12 education: Understanding the status and improving the prospects.* Washington, DC: The National Academies Press. https://doi.org/10.17226/12635.

National Association for the Education of Young Children (NAEYC). (2010). *Early childhood mathematics: Promoting good beginnings.* Retrieved from https://www.naeyc.org/files/naeyc/file/positions/psmath.pdf.

National Association for the Education of Young Children, & National Council of Teachers of Mathematics. (2009). *Where we stand on early childhood mathematics.* Washington, DC: National Association for the Education of Young Children.

National Research Council. (2012). *A framework for K-12 science education: Practices, crosscutting concepts, and core ideas.* Washington, DC: The National Academies Press.

Perry, B., & Dockett, S. (2008). Young children's access to powerful mathematical ideas. In L. D. English (Ed.), *Handbook of international research in mathematics education* (2nd ed., pp. 75–108). NY: Routledge.

Perry, B., & Dockett, S. (2013). Reflecting on young children's mathematics learning. In L. D. English, & J. T. Mulligan (Eds.), *Reconceptualizing early mathematics learning* (pp.149–162). Dordrecht, the Netherlands: Springer.

Purzer, Ş., Goldstein, M. H., Adams, R. S., Xie, C., Nourian, S. (2015). An exploratory study of informed engineering design behaviors associated with scientific explanations. *International Journal of STEM Education, 2*(9). https://doi.org/10.1186/s40594-015-0019-7.

Sarama, J., & Clements, D. H. (2009). *Early childhood mathematics education research: learning trajectories for young children.* New York: Routledge.

Van Meeteren, B., & Zan, B. (2010). Revealing the work of young Engineers in early childhood education. *STEM in Early Education and Development Conference.* Retrieved September 7, 2017 from http://ecrp.uiuc.edu/beyond/seed/zan.html.

Watkins, K., Spencer, J., & Hammer, D. (2014). Examining young students' problem scoping in engineering design. *Journal of Pre-College Engineering Education Research, 4*(1), 43–53. https://doi.org/10.7771/2157-9288.1082.

Zawojewski, J., Bowman, K., & Diefes-Dux, H. A. (Eds.). (2008). *Mathematical modeling in engineering education: Designing experiences for all students.* Roterdam, the Netherlands: Sense Publishers.

Chapter 3
Encouraging the Development of Engineering Habits of Mind in Prekindergarten Learners

Christine N. Lippard, Katie L. Riley and Monica H. Lamm

Abstract Experiences in early childhood set the foundation for lifelong learning. Given the integrative and applied nature of engineering and children's natural curiosity, we suggest that prekindergarten classrooms are well suited for providing opportunities to promote the development of engineering habits of mind (EHM). Developmental theories suggest that children learn best through hands-on experiences that enable them to explore and discover concepts themselves and that others in the child's environment can serve as active partners in exploration. Recognizing the emphasis on integrated curriculum in early childhood and the competing demands for time in preschool classrooms, we identify the EHM as an appropriate early engineering emphasis that can be embedded in everyday classroom moments. To this end, this chapter begins by pointing out connections among science, math, and engineering for early learners, highlights theories that inform our work with engineering in prekindergarten classrooms, discusses EHM in prekindergarten learners, briefly presents a pilot study of observing EHM in prekindergarten classrooms, and ends by drawing overarching conclusions and suggesting future directions for incorporating EHM into prekindergarten classrooms.

3.1 Introduction

Within the prekindergarten environment, children are drawn to opportunities that naturally engage them in engineering processes, skills, and thinking (Bagiati & Evangelou, 2011, Chap. 6; Gold, Elicker, Choi, Anderson, & Brophy, 2015). Recent work from leading psychologists indicates that prekindergarten children are inclined to think like engineers. Children are open to taking in new information and effective at using it to formulate hypotheses, even more so than adults (Lucas, Bridgers,

C. N. Lippard (✉) · K. L. Riley
Department of Human Development and Family Studies, Iowa State University, Ames, IA, USA
e-mail: maynard1@gmail.com

M. H. Lamm
Department of Chemical and Biological Engineering, Iowa State University, Ames, IA, USA

L. English and T. Moore (eds.), *Early Engineering Learning*, Early Mathematics Learning and Development, https://doi.org/10.1007/978-981-10-8621-2_3

Griffiths, & Gopnik, 2014). They are more likely to attempt systems thinking when given open-ended opportunities with materials, as opposed to when they are given direct instruction (Bonawitz, Shafto, Gweon, Goodman, Spelke, & Schulz, 2011). Opportunities to engage in engineering thinking and testing are abundant in prekindergarten classrooms, as several chapters in this book have illustrated. Such opportunities may occur when children run out of a paint color and decide to mix two colors to produce the color they want, use blocks to build a bridge, or investigate how a new toy in the classroom operates. Each of these situations offers children an opportunity to engage in engineering—solving problems and making decisions within a given set of constraints to meet a goal (Katehi, Pearson, & Feder, 2009). These same situations require children to apply practical math and science principles to address their needs or wants.

Prekindergarten children who participate in engineering thinking and learning are better equipped for math and science learning. In terms of early mathematics, children's skills can be classified into those related to quantity and numeracy and those related to geography and spatial thinking (Clements & Samara, 2007). For example, Verdine and colleagues (2014) define spatial skills as "mentally manipulating information about the structure of the shapes and spaces in one's environment "(pg. 1062). They found that prekindergarten children's spatial assembly skills were related to other early mathematical skills including counting and number sequencing. Building and other design-related engineering activities challenge children to strengthen their spatial logic and geometric thinking as they attempt to fit components of a system together (e.g., see Chap. 5). In the prekindergarten classroom, performing an engineering task such as building with Marbleworks® pieces encourages children to explore how pieces of different sizes and shapes can be manipulated and formed to produce a desired structure. Indeed, early engineering play in block building is associated with children's achievement in math courses into middle school (Wolfgang, Stannard, & Jones, 2001). Beyond traditional engineering play of building, addressing engineering problems such as moving water during water play exposes children to math concepts of measurement, volume, and conservation.

Science skills are also utilized in engineering play in the prekindergarten classroom. Hypothesis generating and testing are key science skills children practice in engineering solutions to everyday classroom problems. For instance, if children drop a toy behind a shelf, they have now set up a problem to solve in trying to retrieve the toy. Children may hypothesize that the toy can be retrieved with a ruler, some string, or a long piece of tape and then move to testing each of these. In addition to testing hypotheses, children will learn about the characteristics and limitations of various materials (e.g., a long piece of tape sticks not only to the toy you intend, but also to the wall or itself; rulers are stiff which makes them better than string for pushing toys). As children encounter these problems and make use of various materials, they are likely to encounter science concepts related to textures, mixtures and solutions, density, solubility, geology, heat transfer, and even chemistry. Further illustrations of these various STEM developments appear in Sect. 2.

3.2 Theory

Prekindergarten children (3–5 years) learn through hands-on experiences and interactions with others (Dewey, 1997; Piaget & Cook, 1954; Vygotsky, 1978/1997). This understanding is guided by two theoretical perspectives—constructivism (Dewey, 1997; Piaget & Cook, 1954) and sociocultural theory (Vygotsky, 1978/1997). These theoretical perspectives work in tandem to highlight how children learn when they engage with their environment and others in that environment.

Key propositions we draw from constructivism are that children construct knowledge through manipulating and acting upon materials in their environment and the environment itself (Piaget & Cook, 1954) and that these interactions must be meaningful to the child (Dewey, 1997). Such behaviors as hypothesizing and testing and revising through trial-and-error are indicative of this hands-on learning. Gopnik and Wellman (2012) support the constructivist understanding of learning in their work suggesting that children learn through Bayesian modeling or a series of advanced computations that are made at a subconscious level to determine what the most likely outcome is of a given stimulus.

Sociocultural theory suggests that interpersonal interactions promote new levels of understanding, where continuous interaction with more competent others allows individuals to revise and advance their levels of understanding (Newman & Newman, 2009; Vygotsky, 1978/1997). It is imperative that teachers act as aids and collaborators in the learning process rather than providing direct instruction. This learning approach requires children to be active participants in their learning as they draw on their current skills to help them with the higher-order task. Peers may also take on the role of a more advanced other in the learning process. Like teachers, peers may ask questions or prompt ideas. Mercer and Howe (2012) suggest that sociocultural theory is well suited for explaining teaching and learning because it illuminates both how individuals gain new knowledge (acquiring it from others who possess more than them) and also how the shared knowledge of a group or society progresses through interactions. This is very fitting for learning in engineering as the individual goals children are trying to accomplish in solving problems are situated within a larger context of shared norms and goals.

Constructivist and sociocultural theories have been, at times, pitted against each other. However, the complex and interdisciplinary nature of engineering and the equally complex phenomena of child development necessitate consideration of a multi-theoretical approach. Bruner (1997), while acknowledging distinct differences, suggests integrative points of the two theories. In particular, a socioconstructivist perspective highlights that children learn not only through their own individual interactions with materials and the environment, but also that interactions with other children and teachers may impact on how children interact with materials and the world around them. Further, children learn through interactions and activities that are meaningful to them personally in the context of the meaning held by their larger social context. As we have worked to understand children's development of engineering habits of mind, we are guided by a socioconstructivist perspective that informs our work with three specific expectations: (1) children develop EHM by addressing everyday problems and goals because these problems and challenges are meaningful to them; (2) children learn through actively interacting with and acting on materials in their environment, and (3) children's interactions with materials and the environment can be enhanced by interactions with others (i.e., teachers and peers).

3.3 The Literature Review

Engineering habits of mind are a set of "values, attitudes, and thinking skills associated with engineering" (Katehi et al., 2009, p. 7). Katehi and colleagues define six engineering habits of mind (EHM) to be fostered in K-12 education: systems thinking, creativity, optimism, collaboration, communication, and ethical considerations. The EHM are not a prescribed curriculum, but rather can be viewed as developmental outcomes that arise from children's meaningful interactions with engineering concepts and activities. For this reason, EHM can be embedded and facilitated within existing classroom curricula and practices. Table 3.1 summarizes the six EHM discussed above by listing a definition for each habit and identifying several examples from the K-12 literature that explore that particular habit. It is encouraging that K-12 educators are finding ways to integrate the facilitation and assessment of EHM into existing classroom practice (Besser & Monson, 2014; Bottomley & Parry, 2013; Glancy, Moore, Guzey, Mathis, Tank, & Siverling, 2014; Tank, Moore, Babajide, & Rynearson, 2015; Chap. 4). This is in contrast to facilitation and assessment strategies for EHM that rely on units with particular activities or engineering tasks that must be introduced into the classroom by the teacher, in addition to what is already occurring in the classroom (Chiu & Linn, 2011; DeJaegher, Chiu, Burghardt, Hecht, Malcolm & Pan, 2012; Hobson Foster, Husman & Mendoza, 2013; Loveland & Dunn, 2014). Below we discuss each EHM in further depth and suggest how it might be particularly important for prekindergarten children.

3.3.1 Systems Thinking

Systems thinking facilitates higher-order thinking as children seek to identify and understand interconnectedness and how materials relate to each other and contribute to the system as a whole (NAE & NRC, 2009). The prekindergarten classroom context encourages children's systems thinking by offering opportunities to explore objects within distinct learning areas and to examine interconnectedness when materials are combined across learning areas. These opportunities challenge children to consider the properties and functions of various materials, which also promotes scientific thinking as well as vocabulary development. Rehmann, Rover, Laingen, Mickelson, and Brumm (2011) identify four features that are common to many definitions of systems thinking—(1) "viewing a problem broadly and holistically"; (2) "identify-

Table 3.1 Engineering habits of mind documented in the literature

Engineering habit of mind	Definition	Previous literature exploring the habit
Systems thinking	"involves understanding (1) how individual parts function, (2) how parts relate to each other, and (3) how parts, or combinations of parts contribute to the function of the system as a whole" (NAE & NRC 2009, 122)	Chiu & Linn (2011), Virani, Burnham, Gonzalez, Barua, Fredericksen, & Andrade, (2011), DeJaegher et al. (2012), Bottomley & Parry (2013), Hobson Foster et al. (2013), Moore, Glancy, Tank, Kersten, Stohlman, Ntow, & Smith, (2013a), Moore, Tank, Glancy, Kersten, & Ntow, (2013b), Glancy et al. (2014), Loveland & Dunn (2014), Berry & DeRosa (2015), Tank et al. (2015)
Creativity	"use of imagination in design process … includes originality, flexibility, and imagery" (Loveland & Dunn, 2014)	DeJaegher et al. (2012), Bottomley & Parry (2013), Hobson Foster et al. (2013), Moore et al. (2013a), Moore et al. (2013b), Glancy et al. (2014), Loveland & Dunn (2014), Berry & DeRosa (2015), Tank et al. (2015)
Optimism	"reflects a worldview in which possibilities and opportunities can be found in every challenge …" (NAE & NRC, 2009, p. 152)	Bottomley & Parry (2013), Moore et al. (2013a), Moore et al. (2013b), Glancy et al. (2014), Loveland & Dunn (2014), Tank et al. (2015)
Collaboration	"leverages the perspectives, knowledge, and capabilities of team members to address a design challenge" (NAE & NRC, 2009, p. 152)	Chiu & Linn (2011), Virani et al. (2011), DeJaegher et al. (2012), Bottomley & Parry (2013), Hobson Foster et al. (2013), Moore et al. (2013a), Moore et al. (2013b), Glancy et al. (2014), Loveland & Dunn (2014), Berry & DeRosa (2015), Tank et al. (2015)
Communication	understanding the wants and needs of others and explaining solutions to problems (NAE & NCR, 2009, p. 152)	Chiu & Linn (2011), Bottomley & Parry (2013), Hobson Foster et al. (2013), Moore et al. (2013a), Moore et al. (2013b), Glancy et al. (2014), Loveland & Dunn (2014), Berry & DeRosa (2015), Tank et al. (2015)
Attention to ethical considerations	"draw attention to the impacts of engineering on people and the environment; ethical considerations include possible unintended consequences of a technology, the potential disproportionate advantage" (NAE & NRC, 2009, p. 152)	Hess, Sprowl, Pan, Dyehouse, Wachter, & Strobel, (2012), Bottomley & Parry (2013), Moore et al. (2013a), Moore et al. (2013b), Glancy et al. (2014), Loveland & Dunn (2014), Tank et al. (2015)

ing interdependence and feedback"; (3) "synthesizing as well as analyzing individual components"; and (4) "accounting for dynamic, nonlinear behavior" (p. 3). As mentioned previously, prekindergarten children are still developing their capacity to think abstractly, which is a characteristic of these four features. Thus, concrete forerunner skills of systems thinking must be identified in order to recognize developmentally appropriate expressions of this engineering habit of mind in prekindergarten children. We propose that these include identifying and labeling properties of materials, identifying limits and possibilities of materials, transferring and applying knowledge from one situation to another, and flexible management of materials in ways that promote solving problems in addition to identifying parts of a whole and simple cause and effect within systems.

Young children are capable of systems thinking, and developing this skill enhances their ability to learn in all content areas. In one study, Nelson, O'Neil, and Asher (2008) investigated prekindergarten children learning object labels and found that children better recall object names when the functions of the objects were provided. Further, children inquired about object functions when a new or novel object was named. These findings demonstrate that children naturally look to make connections in their learning, and making meaningful connections is beneficial for children's understanding and remembering new information. This is further a useful habit of mind for children to develop, while they are immersed in integrated curriculum in early childhood education, so that they will develop the ability to self-identify connections as they move into later years of schooling where curricula are often less integrated.

3.3.2 Creativity

Creativity is the use of imagination in solving engineering problems (Loveland & Dunn, 2014). Creativity allows children to thrive in novel situations, encouraging flexibility and the ability to generate a variety of solutions to problems. Creativity also contributes to children's ability to transfer knowledge from one setting to another, which is critical for school readiness because using one's imagination, flexibility, and imagery are skills that help children construct new knowledge in a way that is meaningful to them. Creativity can lead to higher-order thinking as children generate new solutions and ways of thinking to produce an answer to the problem. The development of creativity in the prekindergarten years is considered so important that is included in state early learning standards (Early Childhood Iowa, 2012) and has been associated with higher academic achievement (McCabe, 1991).

3.3.3 Optimism

Optimism reflects the belief that problems and challenges can be viewed as opportunities (NAE & NRC, 2009). Optimism includes children's perseverance and self-motivation, which are in turn important for learning as children who have developed the habit of mind of optimism are motivated to stay engaged in understanding challenging concepts and have developed the perseverance to do so. In prekindergarten, children often exhibit these skills during unstructured play activities. The motivation to continue working on a task even if unsuccessful on the first try and to persevere will be valuable for all learners as they work to master new concepts over time. Optimism can also facilitate positive social/emotional development. How children think about and respond to problems has an impact on how they view themselves, their ability to shape their own learning, and how they handle future problems (Pawlina & Stanford, 2011). This increase in self-efficacy can bolster children's confidence and encourage a growth mind-set (Dweck & Leggett, 1988), which then can act as a cyclical process leading to increased child optimism during the learning process.

3.3.4 Collaboration

Collaboration allows for groups to incorporate the strengths and abilities of each group member into the problem-solving process to reach a better outcome (NAE & NRC, 2009). In prekindergarten, children have opportunities to practice collaborating with peers and teachers in groups, whether they are participating in a large group teacher-led activity or with a few peers in a learning area during free play. Successful collaboration can increase children's understanding of material and help them to engage in deeper thinking about an idea or concept during the learning process as they are exposed to multiple perspectives. From our theoretical perspective, collaboration can be a key driver of development as children's individual schema is challenged. Indeed, collaborative problem solving has been associated with greater learning (Azmitia, 1988). There are indications that collaboration may be particularly beneficial for children with special needs (Colozzi, Ward, & Crotty, 2008) as they are able to observe not only the EHM skills, but also other skills of typically developing peers.

3.3.5 Communication

Communication is an essential skill for collaborative problem solving and ultimately for learning and academic success. Communication in this regard goes beyond vocabulary development, though that will occur, and focuses on the use of communication as a tool. When children have to communicate their ideas, they are challenged to solidify their thinking and in turn expose that thinking to either affirmation or correction by others. Expressing what one is thinking when learning is a way for educators to assess children's understanding and integration of new knowledge. Practicing effective communication in prekindergarten helps bolster language development.

3.3.6 Attention to Ethical Considerations

Attention to ethical considerations emphasizes recognizing that any given solution to a problem will impact others in the environment and the environment itself. An advanced understanding of this would include recognizing the possibility of unintended consequences. Though this EHM may have a social skills component in that emotions may be considered or addressed, attention to ethical considerations really considers the materials-based aspects of a situation. It is possible that attention to ethical considerations, more so than the other habits of mind, will require input from more advanced others. Prekindergarten children are typically capable of following multi-step directions and simple cause-and-effect relationships and are beginning to understand sequencing in activities. These abilities can be scaffolded to help children predict possible outcomes. Thus, as children transition toward more abstract thinking and the ability to see various perspectives, opportunities to attend to ethical considerations provide practice in considering potential consequences of decisions and actions.

3.4 Method for Building a Classroom Observation Protocol

This section describes the process by which an observation protocol for documenting occurrences of engineering habits of mind in prekindergarten children was constructed and piloted. The team assembled to initially build the protocol (Phase 1) consisted of one faculty with domain expertise in early childhood education, one faculty with domain expertise in chemical engineering, and one doctoral student from the human development and family studies program. During pilot testing (Phase 2) of the protocol additional team members were added including a faculty member with expertise in early elementary science and engineering education, another faculty member with expertise in early childhood education, another doctoral student from the human development and family studies program, and two undergraduate engineering students.

To begin the process, for each EHM, we discussed what typical displays of EHM in the classroom would be and what classroom factors would contribute to the occurrence. We generated hypothetical examples from our previous experience with young children. Once we had an initial shared vision for what each EHM could look like in a classroom, we began classroom observations (described below) and spent several iterations discussing, further clarifying our operationalization of each EHM, observing classrooms, and then returning to discussion.

Classroom observations were conducted at mid-morning, when small-group activities and free-play time were scheduled. All observation notes were recorded on paper and recorded in such a way that children's identities remained anonymous. Observation procedures were presented to the university institutional review board (IRB). Phase 1 observations were determined to not be human subject research given that no interactions were occurring between observers and those being observed; Phase 2 observations were deemed exempt on the basis of occurring in an educational setting observing public behavior.

3.4.1 Phase 1: Development and Field Testing

Field testing took place in the laboratory school at a large, Midwestern University in the USA. The two classrooms observed were prekindergarten classrooms with children aged 3 and 4 years. Each classroom had 18 children and 2 teachers.

To establish construct validity, we observed the university laboratory school classrooms and, for each occurrence, we record the EHM displayed, the time duration, the learning setting and focus (literacy, language, math, art, music, motor), the teacher role (active, passive, questioning, directing), and the teacher response. We tried different observation patterns, such as exclusively focusing on one learning center or activity in the classroom for the duration of the observation while capturing in-depth narrative summaries (Richards & Farrell, 2011), observing the classroom at large for extended periods and then coding (Pianta, La Paro, & Hamre, 2008), or observing the classroom for 15 s and recording observations for 15 s in an interval time sampling fashion (Hintze, Volpe, & Shapiro, 2002). During this early phase, we often conferred with each other in the observation booth to discuss specific play moments, child–child interactions, or teacher–child interactions.

3.4.2 Phase 2: Pilot Testing

The pilot study took place in nine early childhood classrooms housed in public elementary schools (n = 5) or Head Start centers (n = 4). Children enrolled in the classrooms ranged in ages from 3 to 5, and each classroom had a minimum of two teachers.

Similar to Phase 1, all observations occurred in the morning and were scheduled during indoor free play and small-group time though a few observations followed children to their outdoor classrooms. For each observation, three or four research assistants observed from inside the classroom. Research assistants observed the same learning area and recorded anecdotal notes for three minutes and then moved on to a different learning area. At the end of 18 min (two 3-min cycles through each of three areas), research assistants coded their notes into specific EHM occurrences. After field testing multiple observation patterns, this pattern was determined to give the best balance of depth and breadth of data. All research notes were recorded in such a way that children's identities remained anonymous and stored in such a way that teachers' identities remain confidential.

Graduate and undergraduate research assistants were trained to conduct the observations. The training module included background on prekindergarten child development, prekindergarten classrooms, and the EHM. Observation, note taking, and coding skills were developed and practiced by viewing videos and making visits to the laboratory school observation booths. Results presented here draw from all of the anecdotal notes of observations recorded by all of the observers.

3.5 Results and Discussion

We are continuing our work to establish the psychometric properties of the EEOT. Here, we share trends drawn from the descriptive results from the body of examples of EHM occurrences observed in real classrooms and highlight opportunities for facilitating EHM. We include anecdotes to help illustrate the possibilities for facilitating EHM that already occur in classrooms. A robust body of real examples from classroom observations is crucial for future efforts aimed at effective teacher professional development, and we continue to work to this end. It is one thing to tell teachers that they can embed engineering in their classroom, but it will be far more effective to show them by giving them examples from real classroom observations.

Table 3.2 shows select observations from the preliminary data collected with the EEOT. One example is presented for each EHM. Teacher facilitation prompts are included. When the prompt was directly observed during data collection, the prompt appears as a direct quotation. For examples, where no teacher response was observed, we have added prompts to suggest how a teacher might have facilitated the EHM occurrence. Below we further discuss trends that arose from our data.

3.5.1 Engineering Play Throughout the Classroom

The first trend we noted was that EHM were demonstrated in a variety of activities. For example, during our preliminary observations, we saw children display optimism as they worked to refine their designs both when building block furniture and when making paper airplanes. In the case of the paper airplanes, a group of children displayed self-motivation (optimism) as they worked to test their designs by flying their airplanes around the classroom, then to refine and improve their work by cutting off the tips and other parts of the planes and then flying them again. They demonstrated collaboration and communication as they discussed how these changes could make their planes fly better. We also saw children demonstrate EHM in activities such as

Table 3.2 Engineering habits of mind in prekindergarten classrooms

Engineering habit of mind (EHM)	Corresponding prekindergarten skills for the EHM	Example of the EHM observed in a prekindergarten classroom	Facilitation from the teacher to promote the EHM[a]
Systems thinking	Identifying and labeling characteristics and properties of materials, identifying limits and possibilities of materials, transferring and applying knowledge, materials management	A child made an envelope with a sheet of paper and a stapler. Inserted small paper pieces, shook envelope (tested), then added more staples, and shook (tested) again	You needed an envelope for your little papers and used staples to fasten the sides of the big sheet. What other materials can we use to fasten the sides of an envelope?
Creativity	Identifying limits and possibilities of materials, identifying a problem and generating solutions, transferring and applying knowledge	A child was trying to make a helicopter with spinning blades out of connecting manipulatives. He was unable to balance it. He grabbed straws from a shelf and attached them to the center to make it taller and tried to spin the blades	It looks like your straw bent when you tried to spin the blades. Is there something else that might work?
Optimism	Motivation to gather information and consider problems, trying alternative solutions to solve a problem	A child tried to latch legs onto a structure by pushing down on a LEGO piece. After examining the structure, the child pushed the LEGO piece from underneath the structure and it latched	You looked for another way to attach that piece and did not give up
Collaboration	Identifying a problem and generating solutions, identifying limits and possibilities of materials, materials management	Two children worked together to hold blocks as they built a garage around a car	"I am pleased you collaborate with each other"

(continued)

Table 3.2 (continued)

Engineering habit of mind (EHM)	Corresponding prekindergarten skills for the EHM	Example of the EHM observed in a prekindergarten classroom	Facilitation from the teacher to promote the EHM[a]
Communication	Identifying and labeling characteristics and properties of materials, identifying limits and possibilities of materials, identifying a problem and generating solutions, transferring and applying knowledge	A child built a bridge with two pillars, and it collapsed. The teacher asked, "What can you do to make it better?" The child rebuilt the bridge with a third pillar and then showed his teacher pictures of how he added a third pillar to his bridge to make it more sturdy	Wow. So you added a pillar to fix it? Did it work?
Attention to ethical considerations	Materials management, identifying limits and possibilities of materials, identifying possible unintended consequences	Three children throwing paper airplanes flew them toward another group of children	"Let's throw away from people. Where can we throw instead?"

[a]Teacher facilitation remarks that were observed appear in quotation marks. All other facilitation remarks are suggested next actions that a teacher might use to promote the EHM

art and dramatic play. In one of our observations, several children were pretending with plastic toy shapes. One child commented that he was going to "fire the house" with his shape. Another child asked back, "Is it fire or water?" A third child chimed in that she was going to "Frozen it!" These communications and collaborations about what elements to apply to the pretend house not only helped children practice vocabulary, but also challenged their thinking. Though we initially began creating the EEOT with the goal of observing EHM in all classroom activities, early data made it clear that the vast majority of EHM occurrences were happening in a few classroom learning centers—dramatic play, art, manipulatives, blocks, and outside play. This is somewhat consistent with Gold and colleagues (2015) findings that children were more engaged in engineering play when large moveable parts were available and in dramatic play.

3.5.2 Teachers' Roles

Second, we noted that teachers often played a limited role in children's activities that involved EHM. Table 3.2 demonstrates a few exceptions to this trend; however, in many cases we saw children go about their activities without teacher engagement or we saw teachers redirect play. Redirections, at times, addressed behavior concerns such as children's voice volume, but at other times drew children's attention to other academic concepts such as literacy (e.g., "Oh! Wow! What letters do you see on that box?"). We saw multiple opportunities where a teacher could have stepped into prompt children's use of EHM. For example, we observed a child removes a block from other children's somewhat elaborate building. Ethical considerations in this context would require the child to recognize that taking the block alters the structure, though he may also recognize that taking the block made his friends mad. However, seeing possible unintended consequences would likely have required scaffolding from a teacher. Additionally, children playing with sand, water, and PVC pipe are likely to recognize their ability to manipulate the path of running water and to quickly make use of the pipe to do so. However, it may take questioning or feedback from a teacher for children to consider how the altered path will impact other aspects of their play (e.g., Does this path cut off water to the plastic turtles the children have stacked in one corner? Does the new path dump water on a sand hill that was intended to be dry land?).

3.6 Conclusions and Future Directions

3.6.1 Everyday Moments in Classrooms and EHM

Certainly, there is a growing recognition of the value of facilitating and encouraging engineering thinking and learning in the early childhood classroom, as the chapters in this book and other research have shown (e.g., Bagiati & Evangelou, 2015; Bairaktarova, Evangelou, Bagiati, & Dobbs-Oates, 2012). However, this enthusiasm for engineering exists within a reality of limited time in classrooms. Often there are competing priorities for time in early childhood classrooms, with teachers recognizing the need to maintain child-directed, open-ended play time while still addressing required literacy activities or other content areas. Given this reality, we advocate that early engineering efforts should focus on children's development of EHM and that this effort should be embedded into everyday moments in early childhood classrooms. Embedding EHM maximizes the use of integrated curriculum and maintains opportunities for child-directed activity in a way that stand-alone engineering curricula cannot. We also propose that helping teachers see opportunities to promote engineering in activities and routines that already occur in the classroom will build on their strengths and be less intimidating than adding a new piece to their existing curricula. In the next chapter, van Meeteren provides further examples of how early engineering experiences can develop the important EHM explored here.

3.6.2 *Teachers Need Support to Draw Out EHM in Everyday Moments*

Though approaching engineering in the classroom from an embedded perspective may set teachers up to be more successful, this alone is not enough. Teachers need support in recognizing and drawing out EHM in everyday moments. Over the past two decades, early childhood has been characterized by an emphasis on early literacy (Zigler & Styfco, 2010). As early literacy was made a top priority for early childhood education, early childhood teachers were trained to recognize, embed, and facilitate early literacy development. In our classroom observations, teachers seem to draw out early literacy concepts almost effortlessly, asking children what letters they saw on a box, if they recognize letters from their own name on a display, and encouraging children's storytelling by asking, "what happened next?" More recently teachers have begun to develop this same type of comfort with early math skills, not only in terms of counting how many blocks they put away, but also estimating and comparing whether one bowl might have more peaches than another, or asking children what other triangle shape items they see throughout the classroom. These types of interactions and questions may appear and even feel natural to teachers now, but it took training and effort for teachers to recognize and develop strategies to facilitate early literacy and math in everyday classroom moments. Over time, teachers have developed working scripts that provide them with go-to questions and comments to highlight literacy and math concepts.

Before developing these strategies, it first took time and training for teachers to recognize early literacy and math skills, which often look different from their 3rd–12th-grade counterparts, and to value these "pre" skills (e.g., print awareness, one-to-one correspondence). Teachers will, similarly, need training and support to recognize and appreciate pre-engineering skills. Commenting on properties of materials (e.g., "this foam block squishes if I step on it, but the wood blocks don't") or choosing to try to solve problems related to materials rather than immediately asking for help may not initially appear to be engineering skills, but these are the very early forerunners on which more advanced habits of mind such as systems thinking and optimism are built. A 2009 policy brief by the National Institute of Early Education Research recognized the ongoing challenge of developing an early education workforce that is well grounded and knowledgeable in STEM education and noted this as a challenge to enhancing children's understanding (Brenneman, Stevenson-Boyd, & Frede, 2009). Work with adults and children in museum settings suggests that adult–child interactions around engineering benefit when adults have received some instruction on engineering concepts (Benjamin, Haden, & Wilkerson, 2010; Haden, Cant, Hoffman, Marcus, Geddes, & Gaskins, 2014). Our future directions include working to further understand the support teachers need in encouraging children's EHM and using the data we have to highlight for teachers the EHM opportunities already occurring in prekindergarten classrooms.

With training and support, teachers can develop comfort and skill levels similar to those they have for literacy and math and come to have working strategies for encouraging and enhancing children's EHM.

3.6.3 Engineering Habits of Mind as Readiness Skills

We anchor our work on early engineering to the engineering habits of mind in order to align early engineering learning and thinking with K-12 expectations while allowing for the flexibility to embed early engineering in existing developmentally appropriate curriculum and everyday classroom moments. Beyond this, we appreciate the EHM to be meaningful habits of mind for all children to develop, regardless of their future careers. The organization *P21: Partnership for 21st Century Learning* (2016), which has partnered with 19 states to develop standards and resources promoting twenty-first century skills, designates critical thinking, communication, collaboration, and creativity and *Learning and Innovation Skills* critical for the academic and career success of children today. Three of the four of these directly relate to EHM, and systems thinking is listed as a subskill under critical thinking, which further affirms the notion that all children will benefit from an intentional focus on the development of EHM in early childhood. Development of the EHM (systems thinking, creativity, optimism, collaboration, communication, and attention to ethical considerations) are key habits of mind that set children up for learning throughout their academic career and for engaging in society as productive and thoughtful citizens. As one of our community partners from phase 2 put it, "Don't we want all children to have these skills?"

References

Azmitia, M. (1988). Peer interaction and problem solving: When are two heads better than one? *Child Development, 59,* 87–96. https://doi.org/10.2307/1130391.

Bagiati, A., & Evangelou, D. (2011). *Starting young: Learning outcomes of a developmentally appropriate PreK engineering curriculum*. Proceedings of the Research in Engineering Education Symposium. Madrid, Spain: Universidad Politecnica de Madres.

Bagiati, A., & Evangelou, D. (2015). Engineering Curriculum in the Preschool Classroom: The Teacher's Experience. *European Early Childhood Education Research Journal, 23,* 112–128. https://doi.org/10.1080/1350293X.2014.991099.

Bairaktarova, D., Evangelou, D., Bagiati, A., & Dobbs-Oates, J. (2012). *The Role of Classroom Artifacts in Developmental Engineering*. Proceedings of the American Society for Engineering Education Annual Conference. San Antonio, Texas: ASEE.

Benjamin, N., Haden, C. A., & Wilkerson, E. (2010). Enhancing building, conversation, and learning through caregiver-child interactions in a children's museum. *Developmental Psychology, 46,* 502–515. https://doi.org/10.1037/a0017822.

Berry, A., & DeRosa, D. (2015). *K-12 teachers as curriculum designers in engineering professional development*. Paper presented at the American Society for Engineering Education National Conference, Seattle, WA.

Besser, D., & Monson, D. (2014). *Engineering in the K-12 classroom.* Paper presented at the American Society for Engineering Education National Conference, Indianapolis, IN.

Bonawitz, E., Shafto, P., Gweon, H., Goodman, N. D., Spelke, E., & Schulz, L. (2011). The double-edge sword of pedagogy: Instruction limits spontaneous exploration and discovery. *Cognition, 120,* 322–330. https://doi.org/10.1016/j.cognition.2010.10.001.

Bottomley, L. & Parry, E A. (2013). *Defining engineering in K-12 in North Carolina*. Paper presented at the American Society for Engineering Education National Conference, Atlanta, GA.

Brenneman, K., Stevenson-Boyd, J., & Frede, E. C. (2009). Math and science in preschool: Policies and practice. *Preschool Policy Brief, 19*. New Brunswick, NJ: National Institute for Early Education Research.

Bruner, J. (1997). Celebrating divergence: Piaget and Vygotsky. *Human Development, 40,* 63–73.

Chiu, J. L. & Linn, M. C. (2011). Knowledge integration and wise engineering. *Journal of Pre-College Engineering Education Research (J-PEER),* 1(1), 1-14. https://doi.org/10.7771/2157-9288.1026.

Clements, D. H., & Samara, J. (2007). Effects of a preschool mathematics curriculum: Summative research on the Building Blocks project. *Journal of Research in Mathematics Education, 38,* 136–163. https://doi.org/10.2307/30034954.

Colozzi, G. A., Ward, L. W., & Crotty, K. E. (2008). Comparison of simultaneous prompting procedure in 1:1 and small group instruction to teach play skills to preschool students with pervasive developmental disorder and developmental disabilities. *Education and Training in Developmental Disabilities, 43,* 226–248.

DeJaegher, C. J., Chiu, J. L., Burghardt, M. D., Hecht, D., Malcolm, P. T. & Pan, E. (2012). *Learning common core math concepts with WISEngineering.* Paper presented at the American Society for Engineering Education National Conference, San Antonio, TX.

Dewey, J. (1997). *Experience and Education*. New York: Touchstone.

Dweck, C. S., & Leggett, E. L. (1988). A social-cognitive approach to motivation and personality. *Psychology Review, 95*(2), 256–273.

Early Childhood Iowa (2012). *Iowa Early Learning Standards.* Retrieved from http://www.state.ia.us/earlychildhood/files/ECI_SITE_FILES/IAearlylearningstandards.pdf.

Glancy, A. W., Moore, T. J., Guzey, S. S., Mathis, C. A., Tank, K. M., & Siverling, E. A. (2014). *Examination of integrated STEM curricula as a means toward quality K-12 engineering education.* Paper presented at the American Society for Engineering Education National Conference, Indianapolis, IN.

Gold, Z. S., Elicker, J., Choi, J. Y., Anderson, T., & Brophy, S. P. (2015). Preschoolers' engineering play behaviors: Differences in gender and play context. *Children, Young, and Environments, 25,* 1–21. https://doi.org/10.7721/chilyoutenvi.25.3.0001.

Gopnik, A., & Wellman, H. M. (2012). Reconstructing constructivism: Causal models, Bayesian learning mechanisms, and the theory. *Psychology Bulletin, 138,* 1085–1108. https://doi.org/10.1037/a0028044.

Haden, C. A., Cant, E. A., Hoffman, P. C., Marcus, M., Geddes, J. R., & Gaskins, S. (2014). Supporting family conversations and children's STEM learning in a children's museum. *Early Childhood Research Quarterly, 29,* 333–344. https://doi.org/10.1016/j.ecresq.2014.04.004.

Hess, J. L., Sprowl, J. E., Pan, R., Dyehouse, M., Wachter, C. A., & Strobel, J. (2012). *Empathy and caring as conceptualized inside and outside of engineering: Extensive literature review and faculty focus group analyses.* Paper presented at the American Society for Engineering Education National Conference, San Antonio, TX.

Hintze, J. M., Volpe, R. J., & Shapiro, E. S. (2002). Best practices in the systematic direct observation of student behavior. *Best Practices in School Psychology, 4,* 993–1006.

Hobson Foster, C., Husman, J., & Mendoza, C. (2013). *Engineering Solar Energy in the Fourth Grade Science Classroom.* Paper presented at the American Society for Engineering Education National Conference, Atlanta, GA.

Katehi, L., Pearson, G. & Feder, M. (2009). The status and nature of K-12 engineering education in the United States, The Bridge: Linking Engineering and Society, *39*(3), 5–10. https://www.nae.edu/Publications/Bridge/16145/16161.aspx.

Loveland & Dunn. (2014, March). *Engineering Habits of Mind in Technology Education*. Paper presented at the annual conference of the International Technology and Engineering Educators Association, Milwaukee, Wisconsin.

Lucas, C. G., Bridgers, S., Griffiths, T. L., & Gopnik, A. (2014). When children are better (or at least more open-minded) learners than adults: Developmental differences in learning the forms of causal relationships. *Cognition, 131,* 284–299. https://doi.org/10.1016/j.cognition.2013.12.010.

McCabe, M. P. (1991). Influence of creativity and intelligence on academic performance. *The Journal of Creative Behavior,25,* 116–122. https://doi.org/10.1002/j.2162-6057.1991.tb01361.x.

Mercer, N., & Howe, C. (2012). Explaining the dialogic processes of teaching and learning: The value and potential of sociocultural theory. *Learning, Culture, & Interaction, 1,* 12–21. https://doi.org/10.1016/j.lcsi.2012.03.001.

Moore, T. J., Glancy, A. W., Tank, K. M., Kersten, J. A., Stohlman, M. S., Ntow, F. D., & Smith, K. A. (2013a). *A framework for implementing quality K-12 engineering education*. Paper presented at the American Society for Engineering Education National Conference, Atlanta, GA.

Moore, T. J., Tank, K. M., Glancy, A. W., Kersten, J. A., & Ntow, F. D. (2013b). *The status of engineering in the current K-12 state science standards*. Paper presented at the American Society for Engineering Education National Conference, Atlanta, GA.

National Academy of Engineering and National Research Council (NAE and NRC). (2009). *Engineering in K-12 Education: Understanding the Status and Improving the Prospects*. Washington, DC: The National Academies Press.

Nelson, D. K., O'Neil, K. A., & Asher, Y. M. (2008). A mutually facilitative relationship between learning names and learning concepts in preschool children: The case of artifacts. *Journal of Cognition & Development, 9*(2), 171–193. https://doi.org/10.1080/15248370802022621.

Newman, B., & Newman, P. (2009). *Development through life: A psychosocial approach*. Belmont, CA: Wadsworth, Cengage Learning.

Pawlina, S., & Stanford, C. (2011). Preschoolers Grow Their Brains: Shifting Mindsets for Greater Resiliency and Better Problem Solving. *Young Children, 66*(5), 30–35.

P21: Partnership for 21st Century Learning. (2016, February 15). http://www.p21.org/index.php.

Piaget, J., Cook, M. (trans). (1954). *The Construction of Reality in the Child*. Abingdon, Oxon: Routledge.

Pianta, R. C., La Paro, K. M., & Hamre, B. K. (2008). *Classroom Assessment Scoring System (CLASS) Manual, Pre-K*. Baltimore, MD: Brookes Publishing.

Rehmann, C. R., Rover, D. T., Laingen, M., Mickelson, S. K. & Brumm, T. J. (2011). *Introducing systems thinking to the engineer of 2020*. Paper presented at the American Society for Engineering Education Conference in Vancouver, British Columbia.

Richards, J. C., & Farrell, T. S. C. (2011). *Classroom observations in teaching practice. Practice Teaching: A Reflective Approach*. Cambridge: Cambridge University Press.

Tank, K. M., Moore, T. J., Babjide, B., & Rynearson, A. M. (2015). *Evidence of students' engineering learning in an elementary classroom.* Paper presented at the American Society for Engineering Education National Conference, Seattle, WA.

Verdine, B. N., Golinkoff, R. M., Hirsh-Pasek, K., Newcombe, N. S., Filipowicz, A. T., & Chang, A. (2014). Deconstructing building blocks: Preschoolers' spatial assembly performance relates to early mathematical skills. *Child Development, 85*(3), 1062–1076. https://doi.org/10.1111/cdev.12165.

Virani, S., Burnham, I. B., Gonzalez, V., Barua, M., Fredericksen, E., & Andrade, S. J. (2011). *Work in progress: Designing an innovative curriculum for engineering in high school.* Paper presented at the American Society for Engineering Education National Conference, Vancouver, BC.

Vygotsky, L. (1997). Interaction between learning and development. *Mind and Society,* pp.79-91. In M. Gauvain & M. Cole (Eds.) *Reading on the Development of Children* (pp.29-35). New York: W.H. Freeman and Company. (Original work published 1978).
Wolfgang, C. H., Stannard, L. L., & Jones, I. (2001). Block play performance among preschoolers as a predictor of later school achievement in mathematics. *Journal of Research in Childhood Education, 15,* 173–180. https://doi.org/10.1080/02568540109594958.
Zigler, E., & Styfco, S. J. (2010). *Hidden History of Head Start*. New York, NY: Oxford University Press.

Chapter 4
The Importance of Developing Engineering Habits of Mind in Early Engineering Education

Beth Dykstra Van Meeteren

Abstract Engineering education in early childhood (preschool through second grade) is a new emphasis in preK-12 education. Many endeavors to introduce engineering to young children use an approach of providing teachers engineering curriculum with prepared lessons or lesson suggestions. An alternative approach is to examine the current educational settings of preschool through second grade to discern existing contexts and activities where engineering is a natural fit. In this chapter, the author invites the reader to examine the high-quality early childhood educational setting and ponder its untapped potential to develop children's engineering habits of mind and the subsequent advantages for children's development.

4.1 Habits of Mind and Early Engineering Education

The *Next Generation Science Standards* (NGSS, 2013) can be a challenge for early childhood teachers. The Standards identifies the science and engineering all K-12 students should know. Teachers examining the document are challenged to include experiences to empower students to meet engineering performance expectations by the end of second grade. Below these performance expectations, the science and engineering practices, engineering core ideas, and the crosscutting concept of structure and function are described in more detail for early childhood educators. Lastly, readers are provided information that includes connections to the standard from other disciplinary core ideas at that grade level, across grade level, and to the Common Core State Standards (CCSS) in mathematics, and English language arts and literacy (NGAC & CCSSO, 2010). It is a lot for teachers to take in. What is left out is how this may look within the early childhood setting. Teachers struggling to find a place for engineering activities within their classroom may find it helpful to consider what

B. D. Van Meeteren (✉)
Iowa Regents' Center for Early Developmental Education, College of Education, University of Northern Iowa, Cedar Falls, IA 506140-0616, USA
e-mail: Beth.vanmeeteren@uni.edu

L. English and T. Moore (eds.), *Early Engineering Learning*, Early Mathematics Learning and Development, https://doi.org/10.1007/978-981-10-8621-2_4

many experts in engineering education believe to be most important in designing high-quality engineering experiences.

A two-year study of the Committee on K-12 engineering education (The Committee), constituted by the National Academy of Engineering and the Board on Science Education at the Center for Education and the National Research Council, examined the scope and nature of existing efforts to teach engineering in K-12 settings. Subsequent findings were released in their report, *Engineering in K-12 Education: Understanding the Status and Improving the Prospects* (Council, 2009). Within the report, the Committee challenged K-12 education to envision the inclusion of engineering education within the American school system. The Council was not content to have K-12 schools teach *about* engineering, but instead advocated for a K-12 engineering education that would *engage* students in engineering and "align with generally accepted ideas about the disciplines and practice of engineering" (Council, 2009, p. 4). To accomplish this, The Committee provided three guiding principles to develop high-quality engineering education. These principles state that engineering experiences from kindergarten through grade 12 should:

(1) emphasize engineering design;
(2) incorporate important and developmentally appropriate mathematics, science, and technology knowledge and skills; and
(3) promote engineering habits of mind. (Council, 2009, pp. 151–152).

The Committee specified that Principle 3, the promotion of engineering habits of mind, was greatest in importance in guiding the development of K-12 engineering education (Council, 2009). Since the publication's release, discussions on engineering education have embraced preschool as part of the K-12 educational system or PK-12.

4.2 Cognitive Demands of Engineering and Their Relation to Executive Functions

The cognitive processes engineers undergo as they design are complex, requiring the ability to (1) focus, (2) consider multiple sets of data, and (3) to flexibly move from analytical to synthetic thinking, or thinking of details within the whole, or the whole in light of the details. These abilities align with what is called executive function skills (EFs). The three core EFs are defined as (1) inhibitory control: self-regulation or self-control—controlling impulsive actions and having the ability to focus; (2) working memory: the ability to hold lots of information or data in their minds to make relationships among elements or concepts; and (3) cognitive flexibility: the ability to take another perspective and not be thrown off by surprises. More complex EFs include problem solving, reasoning, and planning (Blair & Razza, 2007; Diamond & Lee, 2011; Miyake et al., 2000).

Just as complex cognitive processes are essential for successful engineering design, EFs have been found to be essential for academic success, including early learning. A child's high measures of EFs in preschool are related to the child's measures of high ability in math and literacy in later grades and continue to predict success throughout all of a child's schooling (Gathercole, Pickering, Knight, & Stegmann, 2004; Hanline, Milton, & Phelps, 2010). Interestingly, measures of a young child's EFs were more predictive of future literacy success than measures of early phonemic awareness (Blair & Razza, 2007).

Children's EFs can be boosted through specific kinds of activities and classroom environments. Such environments do not expect young children to sit still for long periods of time, but allow for movement within the classroom and time for aerobic exercise. Teachers within these environments refrain from an abundance of direct instruction and instead create experiences that inspire children to ask their own questions and pursue answers (Diamond & Lee, 2011). This type of environment nurtures self-regulation in a positive atmosphere alongside integrative activities that are interesting and challenging (Blair & Razza, 2007). Such environments welcome the idea of engineering activities. When conducted in a developmentally appropriate way, the activities are compatible with and valued for their potential for integration and potential to nurture the development of EFs. This affects the child's success in all domains.

Cognitive processes within the act of engineering have also been described as "engineering habits of mind." The Committee adopted the term "habits of mind" from the American Association for the Advancement of Science in the publication *Science for All Americans* (Rutherford & Ahlgren, 1991). Engineering habits of mind refer to the values, attitudes, and thinking skills related to the act of engineering and align closely with the twenty-first-century skills (Bellanca, 2010). These habits include (1) systems thinking, (2) creativity, (3) optimism, (4) collaboration, (5) communication, and (6) attention to ethical considerations. The Committee provided succinct descriptions illustrating each habit of mind to guide the development of K-12 engineering education (Council, 2009, pp. 4–6). A summary can be seen in Table 4.1.

Reading this summary may lead one to believe engineering is beyond the young child's grasp that young children's instructional time be more suitably devoted to reading and mathematics. However, renowned civil engineer, Henry Petroski (2003), believed these engineering values, attitudes, and thinking skills develop within the years of early childhood. Petroski understood young children inherently engage in engineering through their play. As children play, they devise, invent, and construct their own toys, games, and artifacts that are useful to them. Petroski perceived engineering in children's actions such as moving sand about with dump trucks in a sandbox, building structures out of unit blocks, altering directions in preparing food snacks, or moving objects in references to light sources to create specific kinds of shadows. According to Petroski, the act of design is engrained in children's imagination, in their choices, and in their play activities involving objects (Petroski, 2003). Regrettably, these activities are not recognized or valued within formal schooling and are the first to go when time becomes an issue. What is often viewed and even dismissed as "mere play" is often the beginnings of engineering thinking or habits of mind and deserves protection in the early grades.

Table 4.1 Engineering habits of mind

Term	
Systems thinking	Equips students to recognize essential interconnections in the technological world Appreciates that systems may have unexpected effects that cannot be predicted from the behavior of individual subsystems
Creativity	Inherent (part of the very nature of something, and therefore permanently characteristic of it or necessarily involved in it)
Optimism	A world view in which possibilities and opportunities can be found in every challenge An understanding that every technology can be improved
Collaboration	Leverages the perspectives, knowledge, and capabilities of team members to address a design challenge
Communication	Essential to effective collaboration Essential to understanding the particular wants of a customer Essential to explain and justify the final design solution
Attention to ethical considerations	Draws attention to the impacts of engineering on people Draws attention to the Impact of engineering on the environment Considers possible unintended consequences of technology Considers the potential disproportionate advantages or disadvantages of technology for certain groups or individuals

4.3 The Engineering Connection with Play

Petroski was certainly not the first to recognize the value of play. German educator and creator of kindergarten, Friedrich Froebel (1782–1852), viewed play as the work of the child. He felt educators should capitalize on the productivity of children's play. In 1837, Froebel created a series of six wooden boxes of materials he called a "system of gifts and occupations." These gifts were meant to provide children with focused play experiences. Gifts two through six were called "The Building Gifts" and consisted of various sets of maple wood blocks that included rectangles, cylinders, cubes, and triangles. The blocks were modular and undecorated to encourage discovery and creativity. Froebel believed educational materials should be designed to allow children not only to learn, but to explore and enjoy (Wellhousen & Kieff, 2001).

Many practicing engineers agree with Froebel and Petroski's reverence of children's play with objects and materials. These engineers advocate for educational settings that provide space, time, and materials for activities that allow children to design and build (Martin, 2008; Papert, 1980, 2008; Petroski, 2003; Turkle, 2008; Wright, 1943). Early experiences are essential to develop engineering habits of mind and produce inventors of technologies yet to be imagined. Turkle (2008) documented numerous scientists, engineers, and designers that readily acknowledged their own playful, childhood constructions, and explorations of objects such as blocks and gears. They cited this play as the genesis of their careers in science, engineering, and design; fields that share "a passion for the technical, for formal analysis, for discovery, and for understanding how things work" (Turkle, 2008, p. 7).

Frank Lloyd Wright, perhaps one of the most influential people in twentieth-century design, credits his interest in architecture to a childhood gift his mother purchased for him at the Chicago World Fair: a set of Froebel blocks (Wright, 1943). The geometric shapes and smooth texture of the maple blocks fascinated the young Wright as he used them to construct patterns and structures. Wright often reminisced about these blocks with a sense of fondness, even at the end of his life. The primary forms and figures of these childhood blocks found their way into more than 500 completed works including houses, offices, churches, schools, libraries, bridges, museums, and one golf course (Brewster, 2004).

Seymour Papert (1980) often credited his childhood fascination with gears as the catalyst for a career that led to pioneering artificial intelligence, co-inventing the Logo programming language. Given a toy car operated by gears, the young Papert became enchanted with the toy's gear mechanisms. As a result, he examined and manipulated every kind of gear system at his disposal. Through these multiple experiences with gears, he constructed a deep understanding of how gears worked, mentally turning them in his head and noting the causational relationships of each gear's motion. Papert was so in tune with the workings of gears that he used them as mental models for abstract ideas in mathematics encountered in school. Papert believed educational settings should provide children ample time and space for regular, open-ended encounters with objects and media such as blocks, gears, paints, and markers. According to Papert, early educational environments should be less formal and designed to support "children as they build their own intellectual structures with materials drawn from the surrounding culture (Papert, 1980, p. 32). According to Papert, early childhood encounters with objects and media have a profound effect on a child's intellectual development and life accomplishments. While educational psychologist cannot measure the immediate benefits of these encounters, Papert urged leaders in education to look beyond the short-term and recognize and value these experiences for their long-term benefits (Papert, 1980).

Fred Martin, graduate student of Seymour Papert and now Professor of Computer Science and Director of Student Success at the University of Massachusetts Lowell, wrote that his early childhood experiences with wooden unit blocks, were foundational for his later success in computer science. Unit blocks, invented by Carolyn Pratt over 100 years ago, are a staple in high-quality early childhood classrooms. A unit block is 5.5 inches long, 2.75 inches wide, and 1.275 inches thick. Larger blocks include a double and quadruple unit blocks that are 11–22 inches long. Smaller blocks are fractions of the standard unit block allowing all to fit together in congruence. Martin had the good fortune of attending an experimental school where he had regular access to these wooden unit blocks for long periods of time throughout the primary grades. "I spent days on end building complex roads, garages, and ramps with the blocks. We would test them by driving cars through the worlds we built (Martin, 2008, p. 180). Martin explained that his engagement with materials that had a well-defined and consistent set of first principles that allowed him to build more intricate structures prepared him for BASIC programming language. (Martin, 2008).

He likened the blocks to computer programming in that there is a finite set of *primitives*. The blocks have physical rules in which they connect, and there are standard patterns and ways of using collections of blocks in a coherent way as *that* becomes a building block to be used in larger structures (F. Martin, personal communication, August 27, 2016).

4.4 Existing Early Childhood Contexts Ripe for Early Engineering

Many accomplished engineers credit serendipitous early experiences with objects for their successful careers in design. It is reasonable to assume that engineering habits of mind should not be left to serendipity, but can and should be purposefully planned and nurtured in all children within the school setting. But like the kindergarten teacher at the beginning of the chapter wonders, where does it fit in? The answer can be found within her gaze upon the firehouse block structure in the block center.

Unit Blocks

Just as Martin and Wright found inspiration in wooden blocks, Brophy and Evangelou (2007) found connections to engineering in preschoolers' play with unit blocks. Observing preschoolers within their natural school setting, they found the children were intrinsically drawn to work with these materials at the block center and that the children's own observation and repetitive actions helped build their intuitions of physical properties that governed their intended designs within block building. In short, Brophy and Evangelou came to understand the children were engaging in a rudimentary iterative design process. From time to time, Brophy and Evangelou would engage the child in conversation to determine the child's thinking or plans. The researchers found children were independent in their work and only occasionally asked for an adult's opinion or help. Children self-selected partners or chose to work alone, but the decision to work in a group or alone was not fixed, but dynamic.

Brophy and Evangelou noted that children building structures began to develop a basic understanding of physics and as a result, actually designed with the constraints of physics in mind. One example given was a child who could construct a sophisticated tower that required counterbalancing weight. Upon questioning, the child could predict and explain what would happen if certain blocks were removed from the structure. Another example concerned a child building a roadway. When the child used two blocks of different thickness, he used a ramp to allow the imaginary car to move from one level to the next. Children were merging their imaginary world with the actual world, reflecting their growing understanding of the constraints of physics; the beginnings of engineering thinking. Brophy and Evanglou (2007) observed focused builders; noting the lengthy 40 min, a 4-year-old devoted "to construct, revise, deal with challenges, rebuild as necessary and finally use" a model of a hotel and swimming pools (Brophy & Evangelou, 2007, p. 5). Analyzing the 4-year-old girl's comments and actions, they determined she held a mental model and used this mental model to "correct" children who attempted to assist her. Brophy and Evangelou (2007) called these behaviors precursors to engineering. They viewed block play as a bridge to support young children's developing understanding of how the world works and how children can use their understanding to design and build.

Unfortunately, block play is typically undervalued by teachers and lies on the periphery of the early childhood curriculum. Teachers often view unit block play as a free-choice activity (Casey et al., 2008) despite research showing early block building relating to spatial skills (Brosnan, 1998; Caldera et al., 1999; Kersh, Casey, & Young, 2008; Petersen & Levine, 2014), correlation with later math achievement (Wolfgang, Stannard, & Jones, 2001), and the development children's understanding of space and the physical properties of objects (Kamii, Miyakawa, & Kato, 2004). The kindergarten teacher's problem of figuring out how to fit in engineering in the curriculum could well begin by reconsidering the value of block play.

Unit Blocks and Tracks

Block play had always been highly valued for challenging children's mathematical thinking as well as developing a working knowledge of physics at the Freeburg Early Childhood Program (2001–2007), an experimental early childhood school supported by the Regents' Center for Early Developmental Education at the University of Northern Iowa, Waterloo Community Schools, and Tri-County Head Start. Each of the school's four classrooms serving children ages three through first grade was equipped with a full set of wooden unit blocks. Children had opportunities to build with blocks every day. Their teachers documented the sophistication and processes of building through photographs, video, and child interviews. On faculty inservice days, Director Dr. Rheta De Vries led teachers and researchers in reviewing reviewed this data to discuss the kinds of problems children were posing themselves and how they went about solving them. The team deliberated about things that affected children's creative design and problem solving such as materials, presentation of those materials, the amount of space and time, and adult intervention. One of the materials they added to each classroom's set of unit blocks was tracks to inspire children to explore force and motion. This activity became known as Ramps & Pathways, an activity that was further researched through funding by the National Science Foundation (Counsell et. al., 2016; De Vries & Sales, 2011; Van Meeteren, 2013, 2016; Van Meeteren & Zan, 2010).

Ramps & Pathways is an activity where one-, two-, three-, and four-foot lengths of simple crown molding are introduced as tracks to children with a simple open-ended challenge such as, "What can you do with these materials?" Children are intrigued and begin to move marbles and other objects along the lengths of tracks. Through exploration, the children notice spheres move easily without a child's push or pull when the track is placed on an incline using unit blocks as supports (Counsell et al., 2016; De Vries & Sales, 2011; Van Meeteren, 2016; Van Meeteren & Zan, 2010) (See Photograph 4.1).

Originally viewed solely as a physical science activity to support young children in their investigation of force and motion, the publication of *Engineering in K-12 Education* (Council, 2009) inspired research that determined its value in supporting

Photograph 4.1 Crown molding and wooden unit blocks are used to create an incline

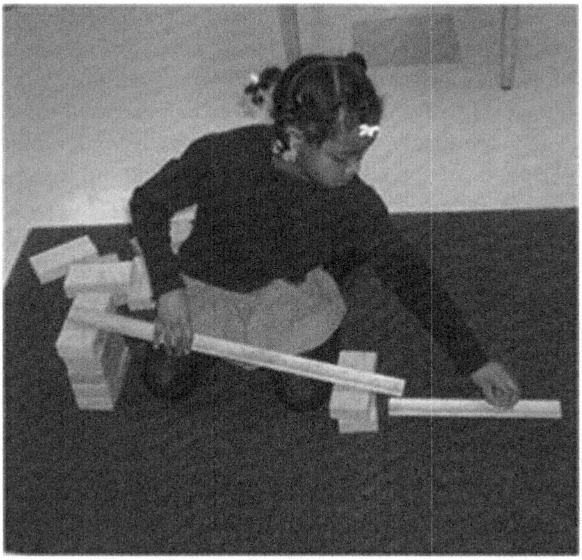

engineering (Van Meeteren, 2013). As children begin to build, they may have a sense of how they want the marble to move and an idea of their constraints, for example the amount of space they have to work within and the number and kinds of blocks available. As children build, they are confronted by constraints of physics and make incremental adjustments or massive design changes to work within those constraints. They begin to think of the whole system as they work with a part or analyze each part of the system while thinking of the whole. They become aware that a change in one part of their ramp system will affect other parts and make decisions of letting some parts of the system go in order to make a more prized part of the system work. (Counsell et., al. 2016; De Vries & Sales, 2011; Van Meeteren, 2013, 2016; Van Meeteren & Zan, 2010). Like block play, Ramps & Pathways engages children in the processes of engineering design and to develop a working understanding that deepens over time.

Even at the very beginning of exploring Ramps & Pathways, children pose their own design problems as they design and build variations of straight pathways using one or multiple sections of crown molding. Children find that decreasing an incline causes a marble to move more slowly and roll a shorter distance, and increasing an incline causes the marble to move faster and farther. They also become aware of differences in the properties of objects they are trying to move and the effect of those properties has on movement. For example, they notice the metal ball bearings are heavier and roll farther off the end of a track than glass marbles that are the same size, but lighter (Van Meeteren, 2016; Van Meeteren & Zan, 2010).

Eventually, children become accomplished with straight pathways and begin to search for ways to get the marble to change direction on a pathway. Through many iterations of trial and error, they figure out that they can change the marble's direction

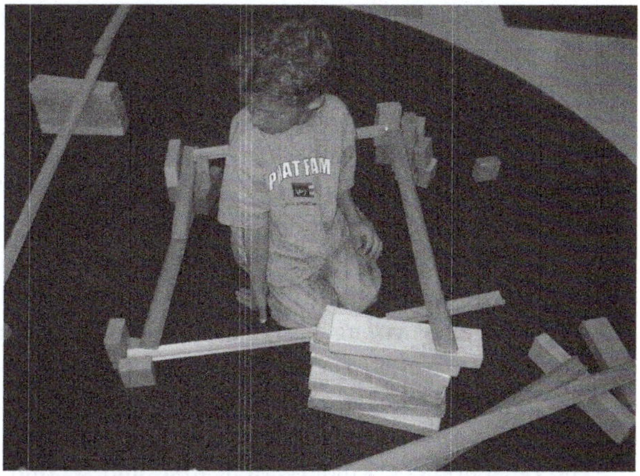

Photograph 4.2 A system with multiple corners

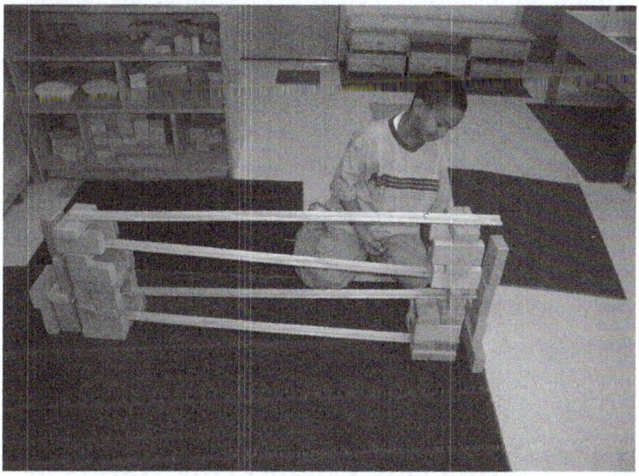

Photograph 4.3 Changing the direction of the marble by alternating inclines stacked vertically

by placing a block in the marble's path. When the marble strikes the block in a particular way, the marble bounces onto the next section of track. Once this is figured out, children challenge themselves to design systems to enable a marble to travel through multiple corners (See Photograph 4.2). Again, children use iterative testing to adjust the inclines to control the speed of the marble; the position of the blocks as supports and barriers; and the position of the receiving track. They persevere until they are successful or until they determine their design is not possible and must be changed (Van Meeteren, 2013, 2016; Van Meeteren & Zan, 2010).

Ever curious about how they can move a sphere in a more interesting way, many children design a system where the tracks can be situated in a vertical position with alternating inclines to allow the marble to drop to the next level and roll in alternating directions (De Vries & Sales, 2011; Van Meeteren, 2016; Van Meeteren & Zan, 2010) (See Photograph 4.3).

These are but a few examples of how three, four, five, six, seven, and eight-year-old children engineer what can be described as marble runs; a type of Rube Goldberg machine.

At the Freeburg Early Childhood Program, Ramps & Pathways was a beloved activity where children returned to it throughout the year, year after year; always with an eagerness to figure out a new design. Teachers have witnessed children grapple with systems that are extremely complex. Systems were designed in a variety of ways to enable the marble to have enough speed to roll through a loop like a roller coaster; down and up multiple hills and valleys; and over wide gaps between two tracks. Children as young as six have designed systems that involved multiple tracks balanced on fulcrums. An eight-year-old built a system that successfully moved a marble 39 feet within a $3' \times 3'$ space. Nearly, all of these designs came from the children themselves.

The preschoolers, kindergarten, first, and second graders in the Freeburg Early Childhood Program were provided time, space, and the materials to build on a daily basis both under teacher observation, and also independently in small groups during center time. Their teachers provided support and encouragement, assisted in documentation, and occasionally fulfilled a child's playful request to assign them a difficult challenge. However, like the children in Brophy and Evangelou's (2007) study, the children typically worked independent of the teacher, posed their own problems to solve, and persevered until they had a working structure or decided their intended design was physically impossible and needed to be reconsidered. Children self-selected partners or chose to work alone, but the decision to work in groups or alone was dynamic (De Vries & Sales, 2011; Van Meeteren, 2013, 2016; Van Meeteren & Zan, 2010).

4.5 Considering Engineering Habits of Mind from the Perspective of the Child

While the children and their teachers may not have realized it, children's actions within block play and Ramps & Pathways are powerful contexts for developing engineering habits of mind. Results of research with children engaging in Ramps & Pathways (Van Meeteren, 2013) found children exhibited developmentally appropriate behaviors compatible with The Council's six engineering habits of mind (Council, 2009).

Engaging in systems thinking. Children engaged in systems thinking as they coordinated a series of components to work together as a whole system. Children found when one component was adjusted; it affected the marble's movement on the following component or components. They were challenged to coordinate their understanding of force and motion with the arrangement of the materials to control the direction and speed of the marble as it traveled from one component to the next. This suggested an awareness and use of cause and effect relationships. Children also developed their ability to operate within the system of the classroom social community. They began to understand the value of cooperative and collaborative relationships and adjusted their behavior accordingly within the social system.

Engaging in creativity. The creativity of children was revealed in their ability to design a system as well as flip, rotate, and repurpose materials in the development of the design. Flexibility in thinking and fluency was further illustrated by children's ability to consider different ways of arranging supports, tracks, or barriers. As children grew in their understanding of how to build supports that were stable, many became more economical in their use of blocks and arranged them in ways that were more aesthetically pleasing. Creativity was found in the originality of their designs. Children did not rely on an adult prototype, but took pleasure in generating their own designs or elaborated on another child's design or prototype. They also demonstrated creativity by resisting premature closure by continuing to add to the complexity of their structure even after a successful test. While teachers could often suggest another challenge of changing the ending or beginning of the structure, the children typically challenged themselves by redesigning the ending, the beginning, or adding another feature to their system such as a jump or a drop.

Developing optimism. Optimism was demonstrated through children's perception of failed tests as valuable information to use in adjustments or redesigns rather than shameful mistakes to be avoided. Children did not abandon their work upon the failure of a test, but persisted by adjusting or rebuilding components. While a few children had the good fortune of being successful on the first try, children made errors that needed to be addressed 84% of the time. The number of times a child failed before succeeding ranged from one to as many as nine.

Engaging in collaboration and communication. Children engaged in the practices of collaboration and communication out of necessity to fulfill their desire to build. Collaboration developed out of a shared interest or a need to figure out how to get a system to work. Children communicated to distribute materials, space, request assistance, offer suggestions, and to share their success. Children often wrote about their experiences or labeled a series of photographs documenting the process of construction.

Considering ethics. While children did not think about worldwide ethical considerations, they did consider the classroom social community when making decisions about work space and use of materials. Analysis of transcripts revealed children developed collaboration and communication skills to coordinate the sharing of workspace and materials within the classroom. Respect for each other in communication and collaboration improved throughout the year. Children stepped through each other's workspace carefully so as not to destroy others' work. There were no intentional knockdowns recorded in early spring, but evidence of children advocating for each other in regard to fair treatment when disagreements occurred.

A comparison between engineering habits of mind as defined by The Council (2009) and operational definitions of children's engineering habits of mind as defined by results of this study can be seen in Table 4.2.

Table 4.2 Engineering habits of mind in young children as they engage in ramps and pathways

Term	NRC (2008) definition of engineering habits of mind	Operational definitions of engineering habits of mind in young children
Systems thinking	Equips students to recognize essential interconnections in the technological world Appreciates that systems may have unexpected effects that cannot be predicted from the behavior of individual subsystems	Adjusts both ends of a track After adjusting one component, coordinates positions of other components before testing
Creativity	Inherent (part of the very nature of something, and therefore permanently characteristic of it or necessarily involved in it)	Generates own design Flips, rotates, or repurposes materials Considers different ways of arranging supports, tracks, barriers Resists premature closure by continuing to add to complexity even after a successful test
Optimism	A world view in which possibilities and opportunities can be found in every challenge An understanding that every technology can be improved	Does not abandon structure after a failed test Uses failed tests as opportunities to find solutions Tries again after multiple failed tests
Collaboration	Leverages the perspectives, knowledge, and capabilities of team members to address a design challenge	Asks for help from a peer Considers suggestions of a peer Asks to test a peer's system Uses peer's system as a model Provides encouragement and/or advice to a peer
Communication	Essential to effective collaboration, Essential to understanding the particular wants of a customer Essential to explain and justify the final design solution	Shares success of structure with peer Explains success of structure with a peer Asks for help and discusses problem with peer Offers advice to a peer Volunteers to build for another and asks what the peer wants in a design Writes or draws about system
Attention to ethical considerations	Draws attention to the impacts of engineering on people Draws attention to the Impact of engineering on the environment Considers possible unintended consequences of technology Considers the potential disproportionate advantages or disadvantages of technology for certain groups or individuals	Coordinates use of space with peers Coordinates use of materials with peers Takes responsibility for knock downs Considers safety

(Van Meeteren, 2013)

4.6 Reexamining Existing Early Childhood Activities for Engineering

Reconsidering engineering habits of mind from the perspective of the child's work in block play and Ramps & Pathways may be valuable in reexamining other existing activities within the child's world for their engineering potential. Appreciating the value of such activities may spur policy makers, administrators, and educators to seek out opportunities to support and include more of these activities within the preschool through second grade school setting. Wright (1943), Papert (1980), and Martin (2008) were fortunate to have had adults that provided them with materials, space and time, and support that sparked imagination, innovation, and invention. While some children have limited access to engineering experiences through before and after school programs such as *First* LEGO League, and summer programs such as Camp Invention, many children are denied access because life circumstances prevent them from attending. Building time, space, and access to open-ended materials in public school early childhood classrooms would allow all children the benefits of developing engineering habits of mind. High-quality preschools and kindergartens already provide materials like blocks, paints, sand and water tables, and dramatic play. However, they are often undervalued for their rich context for children's development of science and engineering practices. When teachers reexamine each area of the classroom and the reflect upon the experiences offered within it, it is helpful to ask themselves, "What is there for children to figure out?" In this act of reflection, the intellectual rigor of the experiences offered in the area, or the lack thereof, reveals itself.

At the art center, children can be provided with choices to paint on horizontal and vertical surfaces. How easy is it to control the paint on a vertical surface? A horizontal? If the paper is glossy or dull, how well does the paint adhere or absorb? Which paper or surface is best for water color or tempera paint? What kinds of effect does using a tapered paint brush have on a painting? A chisel tip brush? A thick or thin brush? Which brush works best with water color? With tempera? As children work within a variety of different constraints, they will notice how properties of materials interact with each other and that they can vary their actions with the materials to produce different results. Over time, they develop preferences for a particular kind of paper, paint, or brush when they have a specific kind of project they want to design. They develop these kinds of preferences because they have learned about the properties of the materials and the possibilities the properties allow. As they gain control over properties, they begin to design and create, making adjustments in design or use of materials when surprises occur. They are engaged in the engineering design process.

At the water table, children can be given transparent and translucent containers, tubes, and funnels with large or small openings; thin or fat; one or more openings. Transparent and translucent allow children to notice the water levels. Which containers hold the most water? Which the least? Which hold the same? Which take longer to fill? To empty? Why does the water come out in a thin stream with this top on? Which are easier to pour from? How can we get water to pour for this into that and then empty into this? As children work with a variety of tubes and containers, they notice how the properties of the tubes and containers affect how the water is contained or released. Over time, they develop preferences for a particular type of container or tube when they want to move the water in a specific way. Again, they develop these preferences because they are learning about the properties of the containers, tubes, and water, and how they can control it. In the act of deciding what to control and how to control it, they are engaged in the engineering design process.

At the dramatic play center, children can be given long bolts of fabric and ribbon in addition to dress up clothes. Which fabrics feel heavy? Light? Silky or slippery? Is there a way to make this rectangle into a square? A square into a triangle? A square into a rectangle? How could I wrap this fabric around my friend to make a gown? A cape? A robe? How could we use this to make a roof? A lake? A river? A river that is flowing fast? Slow? Over time, children have the possibility to notice the weave, texture, and rigidity of fabrics, and how it affects how easily it can be tied, wrapped, folded. They challenge their spatial thinking as they fold to get a desired shape; consider amounts and lengths with what they want to do and how they wrap, fold, and tuck to make a garment. Upon creating the garments and props, they cooperate to create or retell a story. They are engaging in the design process.

4.7 Summary

Technology is defined as "the study of the human-made world, specifically the knowledge, techniques, systems, and artifacts created by humans to satisfy their wants and needs" (Council, 2009, p. 82). Adults are eager to purchase the next iPhone; tablet; car; bike; coffee maker. Young children are as enamored by the human-made world as adults. Just as adults design and create technology to satisfy their wants and needs, young children are constantly designing and creating artifacts to serve their own wants and needs. By this definition, the block structure built to serve the child's wants and needs is technology. A ramp system built to serve a child's wants and needs is technology. A water fountain system built to serve a child's wants and needs is technology. A child's meticulous carving through the mud on a playground to allow the water from one pool to flow to another has created the technology of a canal to serve the child's want and need to move that water from one place to the next. Every day, children design and create sculptures, paintings, songs, games, and processes to satisfy their wants and needs. Embracing this definition reveals that the world of early childhood is a plethora of opportunities to engage children in engineering to develop engineering habits of mind.

Be reexamining existing opportunities for engineering within a developmentally appropriate classroom, teachers have the opportunity to redesign their classroom environment, processes, and procedures to immerse children in the processes of design. Teachers can support the design process by documenting children's efforts to create and construct through interacting and taking photographs, video, or transcripts.

They can use this documentation to revisit the process with the child, inviting the child to comment and critique the work. Displaying the processes and products to invite responses from audiences show the child their work is important, respected, and valued. In such classrooms, engineering is not something children do from time to time. Engineering becomes a way of learning about the world and how it works. Engineering processes become tools to help children cope as well as create and invent; tools that ask children to think flexibly; tools they routinely use. It becomes a way of being. It becomes a habit of mind.

References

Bellanca, J. A. (Ed.). (2010). *21st century skills: Rethinking how students learn*. Bloomington: Solution Tree Press.

Blair, C., & Razza, R. P. (2007). Relating effortful control, executive function, and false belief understanding to emerging math and literacy ability in kindergarten. *Child Development, 78*(2), 647–663.

Brewster, M. (2004). Frank Lloyd Wright: America's architect. *Business Week* (The McGraw-Hill Companies).

Brophy, S., & Evangelou, D. (2007). *Precursors to engineering thinking*. Paper presented at the Annual Conference of the American Society of Engineering Education, Honolulu, HI.

Brosnan, M. J. (1998). Spatial ability in children's play with Lego blocks. *Perceptual and Motor Skills, 87,* 19–28.

Caldera, Y. M., Mc Culp, A., O'Brien, M., Truglio, R. T., Alvarez, M., & Huston, A. C. (1999). Children's play preferences, construction play with blocks, and visual-spatial skills: Are they related? *International Journal of Behavioral Development, 23*(4), 855–872.

Casey, B. M., Andrews, N., Schindler, H., Kersh, J. E., Samper, A., & Copley, J. (2008). The development of spatial skills through interventions involving block building activities. *Cognition and Instruction, 26*(3), 269–309.

Council, T. A. (2009). *Engineering in K-12 education: Understanding the status and improving the prospects*. Washington, DC: National Academies Press.

Counsell, S., Escalada, L., Geiken, R., Sander, M., Uhlenberg, J., VanMeeteren, B., …, Zan, B. (2016). *STEM learning with young children: Inquiry teaching with ramps and pathways*. New York: Teachers College Press.

De Vries, R., & Sales, C. (2011). *Ramps & pathways: A constructivist approach to physics with young children*. Washington, DC: National Association for the Education of Young Children.

Diamond, A., & Lee, K. (2011). Interventions shown to aid executive function development in children 4–12 years old. *Science, 333*(6045), 959–964.

Gathercole, S. E., Pickering, S. J., Knight, C., & Stegmann, Z. (2004). Working memory skills and educational attainment: Evidence from national curriculum assessments at 7 and 14 years of age. *Applied Cognitive Psychology, 18*(1), 1–16.

Hanline, M. F., Milton, S., & Phelps, P. C. (2010). The relationship between preschool block play and reading and math abilities in early elementary school: A longitudinal study of children with and without disabilities. *Early Child Development and Care, 180*(8), 1005–1017.

Kamii, C., Miyakawa, Y., & Kato, Y. (2004). The development of logico-mathematical knowledge in a block-building activity at ages 1–4. *Journal of Research in Childhood Education, 19,* 44–57.

Kersh, J. E., Casey, B., & Young, J. M. (2008). Research on spatial skills and block building in girls and boys: The relationship to later mathematics learning (Chap. 10). In: O. Saracho & B. Spodek (Eds.), *Contemporary perspectives on mathematics in early childhood education* (pp. 233–250). Charlotte, NC: Information Age.

Martin, F. (2008). Basic manual. In S. Turkle (Ed.), *Falling for science* (pp. 268–271). Cambridge, MA: MIT Press.
Miyake, A., Friedman, N. P., Emerson, M. J., Witzki, A. H., Howerter, A., & Wager, T. D. (2000). The unity and diversity of executive functions and their contributions to complex "frontal lobe" tasks: A latent variable analysis. *Cognitive Psychology, 41*(1), 49–100.
National Governors Association Center for Best Practices, & Council of Chief State School Officers (NGAC, & CCSSO). (2010). *Common core state standards*. Washington, DC: NGAC and CCSSO.
NGSS Lead States. (2013). *Next generation science standards: For states, by states*. Washington, DC: National Academies Press.
Papert, S. (1980). *Mindstorms: Children, computers, and powerful ideas*. New York: Basic Books, Inc.
Papert, S. (2008). Objects in mind: Gears. In S. Turkle (Ed.), *Falling for science* (pp. 268–271). Cambridge, MA: MIT Press.
Petersen, L., & Levine, S. (2014). *Early block play predicts conceptual understanding of geometry and mathematical equivalence in elementary school*. SILC Showcase (September). http://bit.ly/1nwuS4Q.
Petroski, H. (2003). Early education. *American Scientist, 91*(3), 206–209.
Rutherford, F. J., & Ahlgren, A. (1991). *Science for all Americans*. Oxford: Oxford University Press.
Turkle, S. (2008). *Falling for science: Objects in mind*. Cambridge, MA: MIT Press.
Van Meeteren, B. (2013). *Designing elementary engineering education from the perspective of the child*. Cedar Falls, IA: University of Northern Iowa.
Van Meeteren, B. (2016). Ramps and pathways and STEM + literacy. *STEM learning with young children: Inquiry teaching with ramps and pathways* (pp. 110–135). New York: Teachers College Press.
Van Meeteren, B., & Zan, B. (2010). Revealing the work of young engineers in early childhood education. *Early Childhood Research and Practice, 12*(2), "Beyond this issue".
Wellhousen, K., & Kieff, J. E. (2001). *A constructivist approach to block play in early childhood*. Boston: Cengage Learning.
Wright, F. L. (1943). *Frank Lloyd Wright: An autobiography*. San Francisco, CA: Pomegranate.
Wolfgang, C. H., Stannard, L. L., & Jones, I. (2001). Block play performance among preschoolers as a predictor of later school achievement in mathematics. *Journal of Research in Childhood Education, 15*, 173–180.

Chapter 5
Spatial Skills Framework for Young Engineers

Lynn McGarvey, Lixin Luo, Zachary Hawes and Spatial Reasoning Study Group

Abstract Engineering is a spatially demanding field. Yet, unlike previously held assumptions, recent research reveals that spatial ability is not innate, but that through experience, education, and intervention, people of all ages can improve their visualization skills—a key component of the engineering habits of mind. In this chapter, we examine the spatial skills that are predictive of success in engineering education and the types of intervention activities that have demonstrated improved performance in STEM subjects. In doing so, we identify parallel assessment measures, skills, and tasks in the development of a spatial skills framework for young engineers that includes physical and mental rotation, symmetry properties, paper folding and unfolding, and cross-cutting objects. These skills are developed through 3D modelling and 2D representational drawings. We illustrate the skills in the framework with samples of work from students in grades 2 and 3 in response to a series of linking-cube tasks in a cube-creature project.

The Spatial Reasoning Study Group is a transdisciplinary team of researchers with expertise in education, psychology, mathematics, and cognitive science. Members contributing to this chapter include Cathy Bruce, Brent Davis, Michelle Arlee Drefs, Krista Francis, David Hallowell, Joanne Mulligan, Joan Moss, Yukari Okamoto, Nathalie Sinclair, Walter Whiteley, and Geoff Woolcott.

L. McGarvey (✉) · L. Luo
University of Alberta, Edmonton, Canada
e-mail: lynn.mcgarvey@ualberta.ca

L. Luo
e-mail: lixin.luo@ualberta.ca

Z. Hawes
University of Toronto, Toronto, Canada
e-mail: zhawes@gmail.com

L. English and T. Moore (eds.), *Early Engineering Learning*, Early Mathematics Learning and Development, https://doi.org/10.1007/978-981-10-8621-2_5

5.1 Introduction

When we think about professionals who need strong spatial abilities, we might think of visual artists such as sculptors, graphic designers, interior designers, photographers, or animators. Architects, pilots, carpenters, and urban planners may also come to mind. Although the role of spatial thinking in these professions appears quite obvious, the need for and reliance on spatial thinking is critical for many other disciplines as well. Indeed, a growing body of research suggests that spatial thinking is a central and uniting feature of science, technology, engineering, and mathematics (STEM) professions. Individuals with strong spatial skills are more likely to enter, enjoy, and succeed in STEM fields (Wai, Lubinski, & Benbow, 2009). More broadly, research has revealed that spatial skills explain academic success in STEM subjects throughout schooling even beyond measures of verbal and quantitative scores (Mix & Cheng, 2012; Webb, Lubinski, & Benbow, 2007).

In particular, engineering is a spatially demanding field. Visualizing components assembled into a unit, imagining what that unit looks like when rotated and viewed from different perspectives, and moving fluidly between 2D graphics and symbolic representations of 3D objects are essential skills for engineers. Spatial skills are needed to solve problems, develop prototype models on paper, design systems in Computer-Assisted Design (CAD) programmes, and communicate ideas (Hsi, Linn, & Bell, 1997).

In past years, the assumption has been that spatial ability is an innate attribute of an individual—a static component of intelligence. However, more recent research confirms that spatial ability is malleable and may be improved given appropriate experiences, opportunities, and practice. Uttal,Meadow, Tipton, Hand, Alden,Warren, & Newcombe, (2013) performed a meta-analysis of 206 spatial training interventions over a 25-year period (1984–2009) and demonstrated that improved spatial performance was possible through various approaches to training, including playing video games, repeatedly completing spatial tests and tasks, and through in-school spatial instruction. Interestingly, even relatively short interventions of just a few hours were shown to boost individuals' spatial skills, with the effects persisting over time. Furthermore, the authors found some evidence that training one spatial skill (e.g. mental rotation) may lead to improvement in other spatial tasks not addressed in training (e.g. mental paper folding).

The importance of spatial ability and its malleability has been a focus of study in post-secondary engineering education for several decades. In 1996, Miller provided a historical review of spatial visualization literature that had been documented in the *Engineering Design Graphics Journal*. This review raised awareness and became a launching point for new research agendas in the field such as: spatial skills needed for engineers; how spatial ability serves as a barrier for entry into and success in post-secondary engineering programmes, particularly for underrepresented groups; and, intervention studies to investigate ways to improve spatial skills, remove obstacles for learners, and improve success in and retention of students (Marunić & Glažar, 2014; Mohler, 2008; Sorby, 2009).

Despite the importance of spatial skills to engineering, mathematics, and science and evidence suggesting that skills are educable, spatial sense receives little attention in early years curriculum (Sinclair & Bruce, 2015; Sarama & Clements, 2009). The National Research Council (2006) has highlighted this as a "major blind spot" in education (p. 6). Since spatial thinking is not an isolated subject with explicit testing, "it often gets lost among reading, mathematics, and all the other content and skills" (Newcombe, 2010, p. 33). Furthermore, spatial thinking (visualizing) is considered one of the important engineering habits of mind (Lucas & Hanson, 2016) that have been highlighted in this section.

There is limited research with regard to what assessment measures are appropriate with young children or the types of experiences that might promote spatial thinking with children. It is interesting to note that in Uttal et al.'s (2013) meta-analysis of spatial skills training, only four of 206 studies focused on children younger than 13 years of age. While there is a dearth of research, recent intervention studies with young children show tremendous promise with regard to developing skills such as mental rotation (e.g. Hawes, Moss, Caswell, & Pliszczuk, 2015).

In this chapter, we describe the development of a framework of essential spatial skills for young engineers. In the development of the framework, we examined the spatial skills assessed in standardized tests in engineering education and the skills emphasized in spatial intervention studies for engineering students. By becoming familiar with the skills and tasks that are predictive of success in engineering, we are able to identify and describe parallel tasks appropriate for children.

Our purpose in developing the framework is to help educators become aware of the importance of spatial skills training, and the kinds of tasks that help develop those skills in young learners. The underlying premise of this project is that educating spatial skills in early childhood has the potential to increase elementary school children's achievement in mathematics and science (Cheng & Mix, 2014; Newcombe, 2010), and doing so may keep the door open for students to choose and be successful in STEM fields later on. We begin our investigation of essential spatial skills by describing spatial thinking.

5.2 Spatial Development

Multiple definitions and descriptors of spatial thinking exist across disciplines and even within disciplines. With no consensus in the literature, numerous overlapping terms have been spawned by researchers, such as spatial perception, spatial relations, spatial orientation, spatial ability, spatial reasoning, spatial sense, spatial cognition, and spatial visualization to name a few. Within the psychological literature, identifying, isolating, and categorizing the components of this cognitive capacity has been a subject of debate, controversy, and confusion for decades (Newcombe & Shipley, 2015).

Rather than attempt to define or tease apart the nuances of spatial terms, we begin this paper by describing Newcombe, Uttal and Sauter's (2013) two broad domains of spatial skills: navigation and tool design. The first domain is the ability to navigate—moving our bodies through space, and understanding the spatial relationship between and among static and dynamic objects. The ability to navigate in our three-dimensional world allows us to locate a place we have never been to before, to recognize that a 2D map represents a 3D space, and to become aware of how landmarks are oriented in space. While all animals are able to navigate using instincts, senses, and landmarks, humans have the additional capability of representing the spatial environment through language and symbol systems. Humans are able to both imagine and depict the surrounding environment, as well as label landmarks and streets. Further, we invent navigational tools such as compasses, sextant, maps, models, and GPS devices. The development of these tools is part of the second human spatial challenge—to design, manipulate, and transform objects and use them as tools. This latter domain is, in many ways, the definition of engineering itself—"design under constraint" (Wulf, 1999); that is, the design of tools, structures, and systems within the associated constraints.

The ability to represent an object and its internal structure gives us the capacity to spatially imagine what an object will look like as it turns and twists or is inverted or reflected. We are able to anticipate the internal structures of an object when slicing through it, decomposing its parts, and by bending, folding, melting, or applying force. Designing and manipulating tools require specific skills such as rotating, cross sectioning, folding, and transforming or what is often referred to as "intrinsic-dynamic skills" (Newcombe & Shipley, 2015; Uttal et al., 2013). It is the intrinsic-dynamic skills that schools of engineering have been assessing as part of entrance exams for over two decades and are the focus of the next section.

5.2.1 Assessing Spatial Abilities in Engineering Education and Early Childhood

Administering standardized spatial ability tests to engineering freshman has been a common practice for some time. The most frequently used and most reliable tests in engineering assess many of the skills associated with tool design or the intrinsic-dynamic skills (Gorska & Sorby, 2008). In particular, they focus on (1) mental rotation; (2) mental paper folding; and (3) cross sectioning. Importantly, as we will describe, research has revealed these spatial skills are highly correlated with success and retention in engineering. In this section, we describe the most common assessment measures used in schools of engineering. In addition, we sought out items assessing the same skill, but suitable for children. Prior to examining the measures associated with mental rotation, mental paper folding, and cross sectioning, we point to a precursor skill that is taken for granted within the adult assessment measures—the translation between 2D representations of 3D objects.

The vast majority of spatial measures currently used are 2D line drawings of 3D objects. In a study with children ages 4–8, Frick and Newcombe (2015) demonstrated

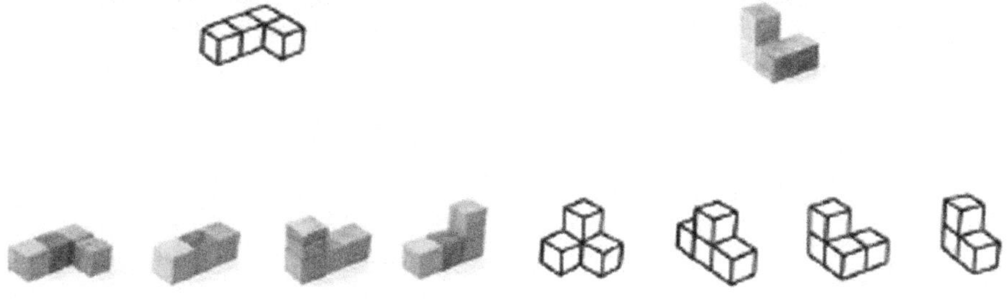

Fig. 5.1 Translation between 2D representations of 3D objects or photographs (Frick & Newcombe, 2015)

that children entering school are capable of recognizing the relationship between 2D and 3D representations of objects but not with one-hundred per cent accuracy. In the study, children were asked to match real 3D objects or photographs of objects with 2D line drawings depicting the schematic edges of objects (and vice versa) (see Fig. 5.1). The results showed that developmentally, by the time children reached 6 years of age and older, their performance became relatively consistent. This is not to suggest that their performance was always correct, but that there was little difference in performance between 6-year-olds and 8-year-olds. This finding suggests that engaging students in tasks involving 2D representations is not inappropriate provided the images are of limited complexity. However, researchers need to use caution when developing spatial assessment measures. Actual objects should be used wherever possible, and if 2D images of 3D objects are used, they need to be relatively realistic (Hoyek, Collet, Fargier, & Guillot, 2012). With this information in mind, we examine the three types of spatial skills most frequently assessed in engineering education.

1. Mental Rotation

Mental rotation is arguably the most extensively studied spatial skill. It involves the ability to look at an object or image and visualize what it will like when rotated in two- or three-dimensional space. There is a close relationship between 3D mental rotation and mathematics (Delgado & Prieto, 2004; Wei, Yuan, Chen, & Zhou, 2012) and other STEM disciplines (Newcombe & Frick, 2010; Wai et al., 2009). Different tests of mental rotation are available, but the two most frequently used with engineering students are the mental rotation test (MRT) from Vandeberg and Kuse (1978) and the Rotations component of the Purdue Spatial Visualizations Test (PSVT:R) from Guay (1977) (see Fig. 5.2).

In both the MRT and the PSVT-R tests, the person has to mentally rotate the given object until a match is made or, in some items, to determine whether another image is the same or a mirror image. The difficulty of mental rotation increases as the angle of rotation from the original position increases, and with whether one or two axes of rotation are at play. Engineering students who perform well on tests of mental

Mental Rotation Test (MRT) (Vandeberg and Kuse, 1978; Peters et al., 1995) *Instructions:* Which two of the four alternatives are rotated images of the first figure?	**Purdue Spatial Visualizations Test – Rotations (PSVT-R)** (Guay, 1976) *Instructions:* Which is the correct answer to the object shown in the middle line when rotated in exactly the same manner as the object in the top?
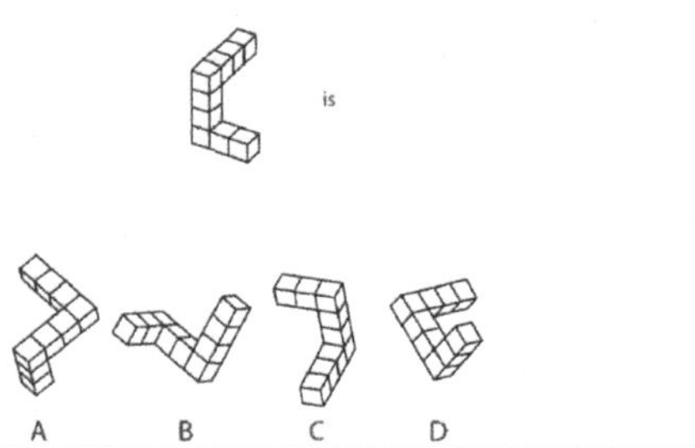	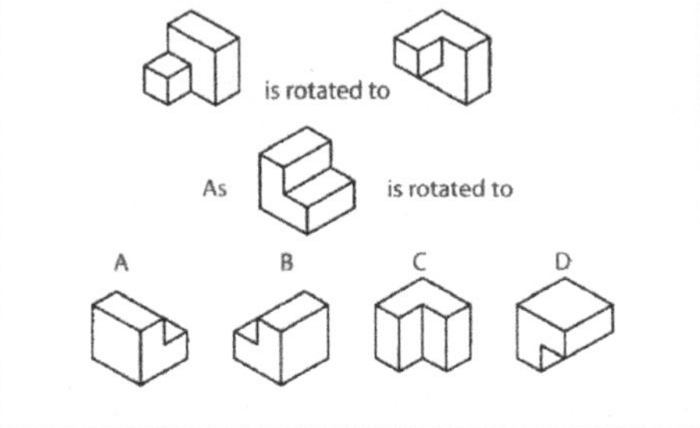

Fig. 5.2 Tests of mental rotation

rotation often perform well in engineering graphics and engineering design courses (Field, 2007; Koch & Sanders, 2011).

Despite the extensive number of studies of mental rotation with adults, relatively few studies—particularly with 3D objects—have been done with young children. Developmentally, the ability to mentally rotate 2D images emerges at about five years of age (Frick, Ferrara, & Newcombe, 2013). In a study using the MRT with children (Hoyek et al., 2012), 7- and 8-year-olds performed at the level of chance. The conclusion was not necessarily that these young children were incapable of mental rotation of 3D objects, but that the measure was likely inappropriate. They suggested that assessment tasks for mental rotation should have fewer choices, simpler instructions, reduction or removal of time constraints, and use of more familiar items. In response to these limitations, researchers have created mental rotation assessments suitable for young children (see Fig. 5.3).

In a spatial skills intervention study with kindergarteners, Casey, Andrews, Schindler, Kersh, Samper, & Copley, (2008) developed a 3D Mental Rotation Task that addresses many of the difficulties identified in Hoyek et al. (2012) study. The 10-item measure uses multilink cubes, rather than images or photographs. The examiner starts by showing the child that the objects are the same by placing them in the same orientation. Then after arranging the two objects behind a screen asks the child how to rotate the second object so that it is, once again, the same as the first. The items increase in difficulty by increasing the number of cubes and complexity of rotations needed.

Hawes, Le Fevre, Xu and Bruce (2015) based their 16-item 3D mental rotation measure on the line drawings of the MRT, but used actual objects. Five or six wooden cubes were glued together. A single blue cube was used as a "mental anchor" to reduce the demands of working memory. The response items included a match, a

3D Mental Rotation Task (Casey et al., 2008) *Instructions:* These two objects are the same. How can you turn one to make them look the same?	**3D Mental Rotation Block Task (3D-MR)** (Hawes, LeFevre, Xu, & Bruce, 2015) *Instructions:* Point to the shape that looks the same as the target item. Show how you can make them look the same.
	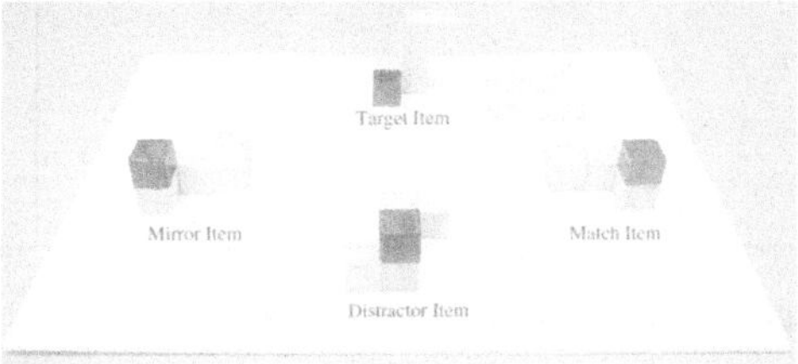

Fig. 5.3 Tests of mental rotation for Children

mirror, and a distractor. The children assessed were told that they were playing a matching game and were asked to point to the perfect match. No time restrictions were imposed. In the study, 29% of 5- to 7-year-olds and 57 per cent of 7- to 8-year-olds were classified as "successful rotators". The measure confirms other studies that suggest that after the age of five there is significant growth in spatial ability, and that there are strong individual differences after this point in development. Given that spatial abilities serve as predictors of science and mathematics achievement (Verdine, Golinkoff, Hirsh-Pasek, & Newcombe, 2017; Wai et al., 2009), these two mental rotation measures not only identify successful rotators, but may also identify young children at risk of not developing the skill.

2. Mental Paper Folding

Another measure of spatial ability associated with success in engineering is mental paper folding. Two different tests of paper folding are used (see Fig. 5.4). The first is the Differential Aptitude Test: Spatial Relations (DAT-SR) (Bennett, Seashore, & Wesman, 1947). This test involves either mentally folding a 2D net or flat pattern into a 3D object or unfolding the 3D representation into a flat pattern. The second test is the Paper Folding and Surface Development Test (Ekstrom, French, & Harman, 1976). In this test, a series of images show how a sheet of paper is folded and punched with one or more holes. The task is to imagine what the sheet of paper will look like when unfolded. Paper folding tests emphasize visualization skills and the tests are predictive of achievement in STEM fields (Hegarty, Kriz, & Cate, 2003; Sanchez & Wiley, 2014; Uttal et al., 2013).

Like mental rotation, mental paper folding is a complex spatial task that is a measure of dynamic spatial transformation. Both skills can be trained and the effects are durable over time; however, there are several notable differences. First, mental rotation involves rigid transformation of an object; that is, the object that is mentally rotated does not change its shape or structure (Harris, Hirsh-Pasek, & Newcombe,

Differential Aptitude Test: Spatial Relations (DAT-SR) (Bennett, Seashore & Wesman, 1947) *Instructions*: Choose the 3D object that would result from folding the first figure.	**Paper Folding Test (PFT)** (Ekstrom, French & Harman, 1976) *Instructions*: Which of the diagrams shows how the paper will look like when it is unfolded?
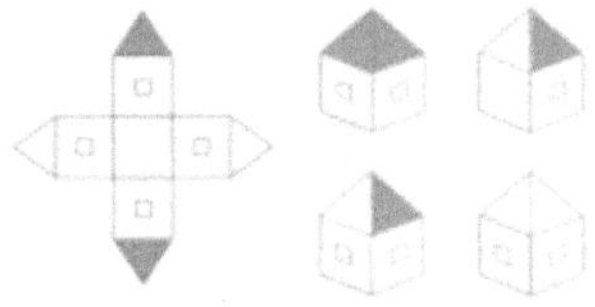	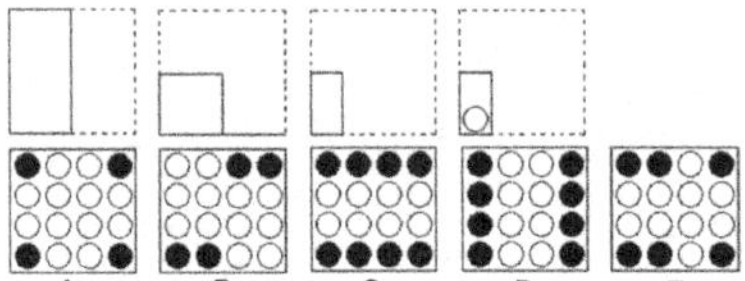

Fig. 5.4 Test of mental paper folding

2013a). Rigid transformation may take place in a 2D or 3D space. On the other hand, paper folding is a non-rigid transformation because the folding and unfolding change its shape. Further, paper folding involves both 2D and 3D aspects for every task. In the DAT-SR example in Fig. 5.4a, the object is transformed from 2D to 3D. In the paper folding test (Fig. 5.4b), the object begins and ends as a 2D form, but the act of unfolding requires a 3D transformation.

Another notable difference between mental rotation and paper folding measures is that mental rotation shows strong differences across genders while paper folding does not (Harris et al., 2013a). While this difference is not well understood, it suggests that the underlying spatial skills are somewhat different.

Despite its noted differences with mental rotation, there are far fewer studies involving paper folding and fewer assessment measures available. Noting this omission, Harris, Newcombe, and Hirsh-Pasek (2013b) developed the mental folding test for children (MFTC) suitable for children ages 4–7 years of age (see Fig. 5.5). Children are shown a piece of paper that is green on the front and purple on the back. For each prompt, they are asked to imagine what the paper will look like when it is folded. Another version of a paper folding task is part of the Cognitive Abilities Test (CogAT-Nonverbal Battery) (2011). Similar to the PFT (above), a page is folded, punched with one or more holes, and the child is to select the image that matches the unfolded paper. This version of paper folding is used in grades 3–12. Levine, Ping, Young, and Ratliff (unpublished) developed a similar version that is valid for children ages 5–10 years of age; however, to reduce the complexity, in their version, the examiner folds and hole punches the sheet of paper in front of the child.

3. Cross Sectioning

The third set of spatial skills we describe is cross sectioning or mental cutting. In these visualization tasks, participants are shown an image of a 3D object and they are to select the 2D image that results from slicing or cutting the 3D representation along a given plane. The Mental Cutting Test (MCT) was originally developed as a subset of a college entrance exam (CEEB, 1939), but has been used widely as a measure of spatial ability of engineering students (see Fig. 5.6). The ability to model

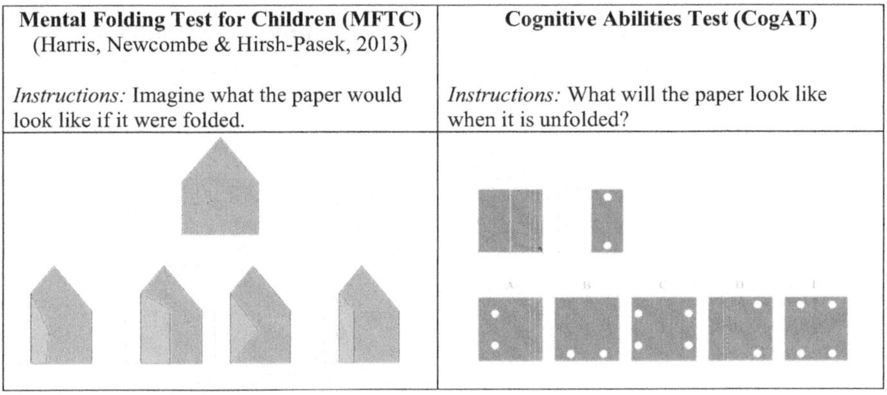

Fig. 5.5 Test of mental paper folding for children

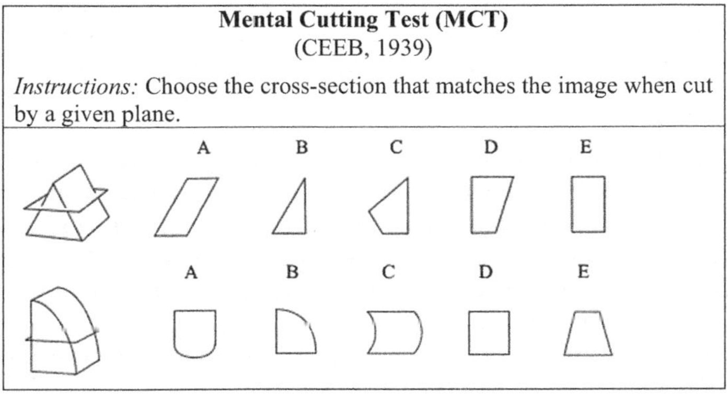

Fig. 5.6 Test of cross sectioning

and manipulate objects in 3D is seen as a necessary skill in descriptive geometry and engineering graphics courses (Tsutsumi, 2005).

While the MCT has been used for 75 years, the relationship between spatial ability and cross sectioning is not well studied. This is at least in part because little is known about how or when the skill develops. In an attempt to study the development of the skill, Ratliff, McGinnis, and Levine (2010) developed a cross- sectioning assessment measure appropriate for children (see Fig. 5.7). The items are based on familiar 3D shapes (e.g. cone, cylinder, pyramid), made from coloured foam, and presented in physical form or as photographs. The objects and photographs of the objects (rather than line drawings) reduce possible ambiguity about the shapes presented. The results indicated that children could successfully complete the task in both 3D and 2D forms. As with the others tasks, performance improved remarkably between the ages 5 and 8. By age 8, the percentage of correct responses was, on average, 88%. However, the authors noted that understanding the individual differences required more study.

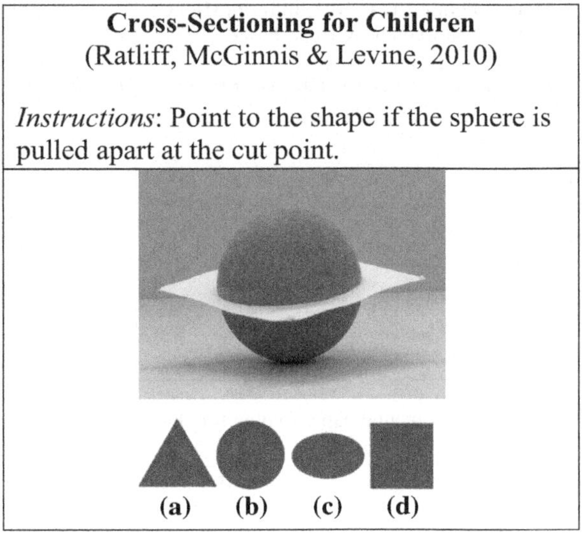

Fig. 5.7 Test of cross sectioning for children

From our review of the literature on spatial assessment measures, we identified that the three spatial skills that are frequently tested and appear to be reliable measures of success in engineering include mental rotation, mental paper folding, and cross sectioning. We also reviewed that children as young as six years of age demonstrate success in child-friendly versions of spatial tasks that use real objects, familiar contexts, simple instructions, limited response choices, and less weight on timed responses. Identifying these skills in the literature provided our first layer of analysis in developing a spatial skills task framework for young learners.

Given that spatial skills can be learned, our next layer of analysis is to determine what tasks and experiences improve individual spatial performance. We now turn to the literature in engineering education where a considerable amount of research has taken place in developing and measuring the types of intervention activities that have been shown to increase spatial skills and overall success with engineering students.

5.3 Spatial Skills Intervention Activities for Engineering Students

Studies examining the effects of training modules, credit courses, intervention programmes, extracurricular activities, and lesson-based materials have contributed to our knowledge of what activities improve spatial skills and also the impact of that improvement on achievement (Branoff, Hartman, & Wiebe, 2002; Sorby & Baartmans, 2000; Ferguson, Ball, McDaniel, & Anderson, 2008; Onyancha, Derov, & Kinsey, 2009). In this section, we focus specifically on a highly cited and successful intervention research programme with engineering students developed

Table 5.1 Spatial intervention course for engineering students (Sorby, Casey, Veurink, & Dulaney, 2013, p. 23)

1. *Surfaces and Solids of Revolution.* 3D surfaces and solids are formed by revolving 2D shapes around an axis
2. *Combining Objects.* New objects are formed by cutting, joining, or intersecting two objects
3. *Isometric Sketching.* Coded plans of cube structures are defined and then sketched from different views
4. *Orthographic Projection with Normal Surfaces.* Objects are represented based on the top, front, and right side views
5. *Orthographic Projection with Inclined and Curved Surfaces.* More complex objects with curves and inclines are represented with top, front, and right side views
6. *Flat Pattern Developments.* 3D objects are represented by folding up of 2D flat patterns
7. *Rotation of Objects about One Axis.* Mental rotation and sketch of an object about an axis
8. *Rotation of Objects about Two or More Axes.* Mental rotation and sketch of an object about two or more axes
9. *Object Reflection and Symmetry.* Symmetric 3D objects are created by reflection and rotation
10. *Cross sections.* The resulting 2D surface from slicing 3D objects an imaginary plane

initially by Sorby and Baartmans (1996), refined over the past two decades and expanded to involve students from eighth grade (Hungwe, Sorby, Drummer, & Molzon, 2007) to post-secondary (Lieu & Sorby, 2015).

The initial project was to develop a training course to improve the visual-spatial skills of first-year engineering students. The programme has been highly successful and multiple studies using this programme demonstrated that post-secondary students who entered engineering with poor performance on a standardized test for spatial ability (e.g. PSVT:R) and who subsequently participated in the training course, showed significant gains in spatial skills; further, the retention rate and GPA of participating freshman on courses in mathematics, chemistry, computer graphics, and design were comparatively higher than for students who performed poorly on the test, but did not participate in the intervention. At present, the spatial intervention course consists of 10 modules taught over a period of 15 h (see Table 5.1).

There are important similarities between the activities within the ten modules and the three types of tasks we identified as key spatial measures including, mental rotation, paper folding, and cross sectioning. Mental rotation of objects is a significant component of several modules including surfaces and solids of revolution (module 1), rotation of objects about one or more axes (modules 7 and 8), and object reflection and symmetry (module 9). Mental rotation is also necessary for success in drawing objects from multiple perspectives (modules 3, 4, and 5). The other two spatial measures, paper folding and cross sectioning, are also explicitly identified in modules 6 and 10, respectively. In addition to these three spatial skills, another skill essential for engineering, specifically tool design, is to be able to visualize how two solids can be combined by joining, intersecting, or cutting one object into another (module 2 and see Fig. 5.8).

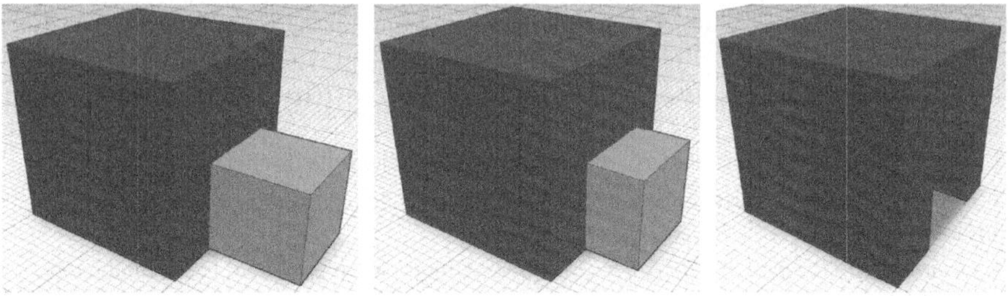

Fig. 5.8 Combining by joining, intersecting, and cutting

The spatial skills of rotation, folding, cross sectioning, and combining form the basis of the tasks across the ten modules. We also note that all of the modules involve 3D models and modelling, and also 2D representations of the 3D objects through sketching or computer-aided design (CAD) drawing.

The emphasis on 3D models as the objects to improve spatial visualization seems obvious from an engineering perspective; yet, it stands in contrast with many school-based activities that are substantially constructed on Euclidean geometry principles with 2D shapes used to understand points, lines, and planes, before turning to three-dimensional objects and spaces. Geometry learning outcomes in primary schools have traditionally included the sorting and labelling of 2D shapes such as squares, rectangles, circles, and triangles, before moving to prisms and pyramids in upper elementary. The emphasis on 2D shapes is markedly different from early childhood programs based on the work of Froebel and Montessori, for example, that begin with the tangible and three-dimensional object (Sinclair & Bruce, 2015).

Finally, a theme across all modules involves the fluid movement between 2D representations and 3D objects through drawings from multiple perspectives. Specifically, 2D representations of 3D objects are part of isometric sketching (module 3), orthographic projections with normal (module 4), and inclined and curved surfaces (module 5). Representing 3D objects in two dimensions is often an overlooked skill; yet, one that proves challenging for many people. Reading blueprints and maps, and building furniture or household items from a set of instructions and 2D images all assume that people are able to translate easily between two and three dimensions. Sorby (2009) concludes that sketching 3D objects contributes substantially to spatial development. The physical act of putting a drawing hand to paper is seen as more effective, at least initially, than creating representations through CAD drawings to develop spatial skills.

We concluded our analysis of the spatial assessment measures and intervention studies with the skills and processes that contribute to success in engineering education. In particular, we noted that the spatial skills involved both rigid transformations through rotation and symmetry, and non-rigid transformation of paper folding, and cross sectioning. These skills were situated within an environment of 3D modelling and representing 3D objects with multi-perspective 2D sketches. This analysis formed the basis of our spatial skills framework for young engineers.

5.4 Spatial Skills Framework for Young Engineers

While there is still much to learn, the studies assessing spatial skills with children suggest that rapid development of spatial skills occurs between the ages of 5 and 8 and that even with that growth there are significant individual differences. Further, evidence suggests that the development of spatial skills has its roots in activities and experiences in childhood. A few examples include, playing with construction toys, particularly those that require following instructions for building (Sorby & Baartmans, 2000), playing 3D computer games (Feng, Spence, & Pratt, 2007; Sorby & Veurink, 2010), playing sports requiring high levels of hand-eye coordination (Lord & Garrison, 1998), and sketching and working with hand-held models (Sorby, 2009). Wai et al.'s (2009) study found that youth with strong spatial skills also showed strong interest in working with hands-on tasks such as building, modelling, repairing, and manipulating tangible objects, such as taking apart toys or electronics and putting them back together. While we can see aspects of the assessed skills identified previously in these activities, we recognize that the studies are correlational. That is, children and youth with strong spatial skills also tend to be more likely to engage in building, modelling, and drawing activities. Other researchers have also emphasized spatial skill development through activities such as copying, drawing, and block building (Casey et al., 2008; Tzuriel & Egozi, 2010). It is a relatively small body of literature that investigates what activities and experiences might serve to develop spatial thinking more directly with children.

Our goal for this paper was to identify the essential spatial skills underlying success in engineering education and translate those skills into a framework for developing spatial skills in early childhood. We wanted to ensure that the skills were relevant for the discipline of engineering and also appropriate for young children. Through our analysis, we identified the following features: mental rotation, paper folding, cross sectioning, and combining objects. These skills are necessarily developed through 3D modelling and through 2D representational drawings from multiple perspectives. From this analysis, we developed a framework for spatial thinking skill developed (see Fig. 5.9). We see the framework as a touch point for early educators in developing lessons and resources for young engineers.

In this section, we provide a description of the different parts of the framework with a set of connected tasks from a cube-creature project (see Appendix A). The tasks in the project were based on the work of Moss, Bruce, Caswell, Flynn, and Hawes (2016) who field-tested research-based activities to develop young learners' spatial reasoning. Our intention is to illustrate the framework using tasks from a cube-creature project along with samples of work from students in grades 2 and 3 collected in four 45–60 min sessions over a four-week period. We begin by describing the initial cube-creature activity in which students create linking-cube structures based on a set of constraints. We then provide examples of tasks to emphasize physical rotation, mental rotation, and paper folding with an emphasis on 2D representations of the creatures created in multiple forms (e.g. coded plans and orthographic projections) and from multiple perspectives.

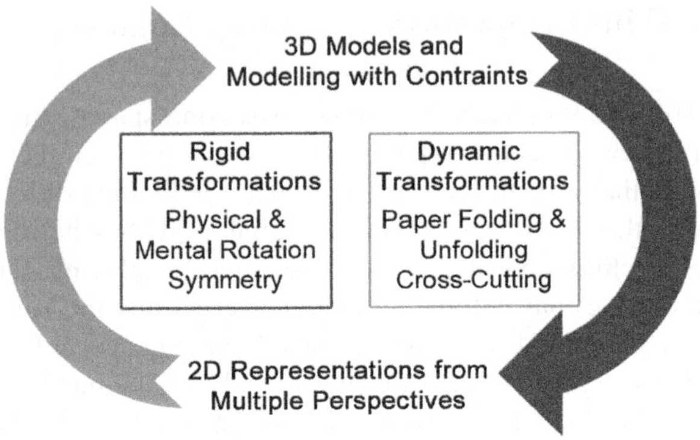

Fig. 5.9 Spatial skills framework for young engineers

Cube-Creature Construction

Use linking cubes to construct Cube-Creatures based on the following constraints:

- Creatures are only one colour
- Creatures consist of exactly 6 cubes
- Creatures stand on four cube-feet. No other body part touches the surface.
- Creatures stand on their own and stay connected when picked up.
- Creatures have no overhanging cubes.

Fig. 5.10 Constraints for creating cube-creature structures

5.4.1 3D Models and Modelling with Constraints

As noted previously, the assessment measures and training tasks consistently involved 3D models and modelling. Although building with blocks is a mainstay of early childhood classrooms, multiple studies with young children demonstrate that guided or problem-solving tasks with blocks are typically more beneficial for developing spatial skills than free play (Bagiati & Evangelou, 2016; Casey et al., 2008; Kersh, Casey, & Mercer Young, 2008). Goal-oriented design and modelling with constraints also more closely resembles the work of design engineers.

To illustrate modelling with constraints using tangible objects, we developed and extended a task in Moss et al. (2016) in which students created "creatures" using linking cubes. Figure 5.10 illustrates the design instructions for creating the creatures.

The constraints were intended to reduce distraction (one colour only), reduce complexity for drawings (no overhanging cubes), ensure ease of handling (stay upright and connected), and allow for a large number of possible structures (6 cubes and 4 cube feet). The task was given to students in grades 2 and 3. We illustrate the work of four students, Jonas, Eli, Nevah, and Marlee, here and throughout the description of the tasks within the framework.

The students were initially asked to create at least three different cube-creatures based on the constraints. The students corrected their own structures and the structures

Fig. 5.11 Jonas's three cube-creatures

of their classmates as they negotiated the constraints. Figure 5.11 illustrates three different structures fitting the given constraints that were created by Jonas.

Altogether the group of four students created twelve structures, but as we had anticipated, there were several duplicate structures. The next task posed to the students illustrates rigid transformations through physical and mental rotation within the spatial skills framework.

5.4.2 *Rigid Transformation: Rotation and Symmetry*

Rigid transformations preserve the shape of an image while it undergoes translation, reflection, or rotation. Visualizing how objects look after they are moved, flipped, or turned is essential for engineers. Our analysis of the research revealed most of the literature in spatial thinking has focused on and emphasized mental rotation. It is clear that young children are capable of such transformations, and that the skill can be improved with training, but children are more successful when tangible objects are provided and the instructions are simple (Hawes, Moss, Caswell, & Poliszczuk, 2015). In the mental rotation assessment tasks, we described previously, participants are asked to match a target object with the same object presented from a rotated perspective. Finding matches of cube structures was the basis of mental rotation tasks for young children in Casey et al. (2008) and Bruce and Hawes (2015). In a similar way, in the cube-creature set of activities students continued to build creatures but they pooled their structures together to look for matches and to describe how the structures were the same and how they were different. The task was extended by asking the group of students to classify and sort their cube-creatures into groups, and to continue building new creatures that fit the original constraints plus the constraints imposed through classification.

The discussion of the four grade 2 and 3 students introduced earlier noted that the structures they created were "2-tall" and "3-tall". When asked if other heights were possible, such as 1-tall or 4-tall, they experimented and explained that these were not possible given the constraints. Nevah said, "It cannot be one [tall] because you can only have four down" or four cubes touching the table top. After several attempts at

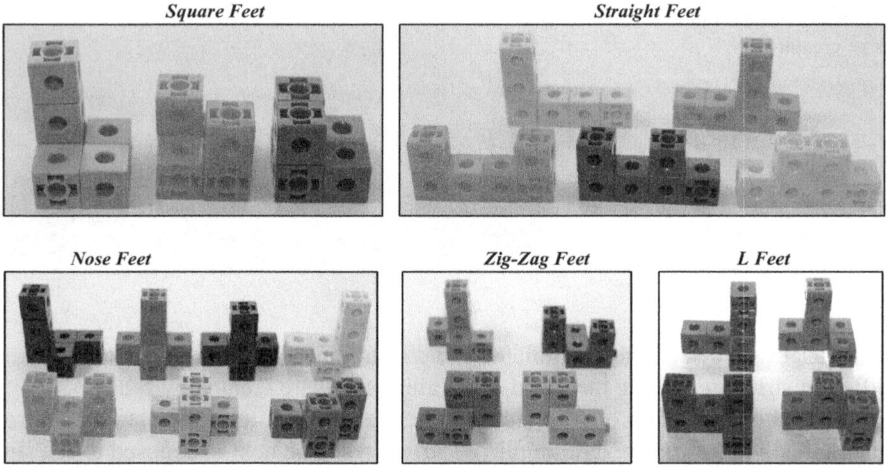

Fig. 5.12 Classification of cube-creatures

Fig. 5.13 "Opposite" figures

creating a creature that was 4-tall, Eli reasoned with four cubes as the base, "There are only two [cubes] left". Therefore, "It can't go higher" than three cubes.

When asked about other ways the creatures were the same or different, they discussed the arrangement of the four "feet" cubes. Jonas, Eli, Nevah, and Marlee, as did many other groups of students who engaged with this task, began to sort, classify, and compose new creatures based on the arrangement of the bottom four cubes. By doing so, they eventually generated five categories of creatures based on their "feet": square, nose, straight, zigzag, and L feet (see Fig. 5.12).

Through this process, the students noted how some variations of the square and line feet creatures looked the same from the front and from the back (i.e. rotational symmetry). They also recognized that some structures looked the same but were "flipped" or "opposite" (see Fig. 5.11). For example, the students were challenged to determine if the blue and green creatures shown in Fig. 5.13 were the same or different. At first, they said they were the same, but when put side by side, Eli said, "this one has this foot ahead" and the other one is "opposite". He tried rotating one around to make it the same, but it was always opposite. The students briefly discussed whether they should make a new category, but chose not to.

By the end of the session, the students had created 23 different creatures classified into five different subgroups. They were confident that they had found all of the different types of foot patterns. They were also sure that there were only three different creatures with square feet. However, they knew there were more cube-creature structures that could be created within the other foot patterns (e.g., zig-zag feet). The teacher–researcher also played a key role in pointing to and asking questions that helped students attend to spatial aspects such as plane symmetry (e.g. "It looks the same on both sides"), rotational symmetry (e.g. "It looks the same when you turn it around"), mental rotation (e.g. "Are those two the same? Can you tell without touching them?"), and visualization (e.g. "What block could you move to make it different?").

5.4.3 Dynamic Transformation: Paper Folding and Unfolding

Unlike rigid transformations that maintain the original dimensions of an object, dynamic transformations involve transforming a 2D object into a 3D structure. In the spatial assessments and spatial training, flat patterns involving real or imagined paper folding has been the primary vehicle for training skills in dynamic transformation. In assessment and spatial training tasks with children, origami, pop-up paper engineering, and building paper airplanes have also been used to develop dynamic transformation skills (Harris et al., 2013b; Taylor & Hutton, 2013).

In the spatial activities for primary-aged students, we chose to develop dynamic transformation skills through the task of creating rectangular prisms by creating and folding paper nets. The students were asked to pick and name one creature from the collection. As they made their selections, the teacher–researcher encouraged them to choose creatures from different categories. To help them remember the face and the feet of their creature (in case it got knocked over), they were to put a yellow eyeball sticker on the front and a blue sticker on the bottom. The stickers served as important anchors and orientation points throughout the remainder of the project (see Fig. 5.14).

Students were asked to create a 2D plan for a cage that would snugly fit their selected cube-creature. To facilitate the task, students were given pre-cut paper of different dimensions and were prompted to select the floor, four walls, and a roof from the paper. The focus was to visualize and construct the size and possible flat patterns that could fold into rectangular prisms that could enclose their creatures.

The students were prompted to begin by choosing a rectangle that would be used for the cage floor and would fit their creature's feet. Then, there were encouraged to view their creatures from the top, and then from the front, back, and sides with their eyes at table height. Students laid out the pieces for their cage with their creature in the middle and were asked to imagine folding up the sides of the cage. Once they were confident that the pieces were oriented correctly, they taped the pieces together into

Fig. 5.14 Student-selected cube-creatures

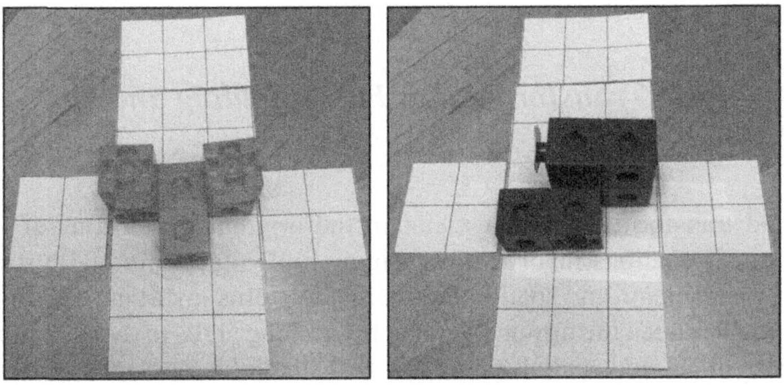

Fig. 5.15 Cages that fit other creatures

a flat pattern, and attached the roof to the back wall (similar to a box and lid shown as an example). The teacher–researcher asked a number of questions to prompt spatial reasoning such as, "Will any of the other creatures fit in your cage?" Students were then encouraged to examine the collection of creatures they had created initially and select ones they thought would fit. The students used both visualization and trial and error through rotation to find possibilities. Using the selected creatures chosen for cage building, Nevah and Marlee noted that they had built the same cage (i.e. $3 \times 2 \times 3$) and Eli noticed that the cage he built for his creature would also contain the creature that Jonas created although the creature had to "turn sideways" (see Fig. 5.15).

Once students had laid out their 2D flat pattern, they taped and folded them into a 3D cage (see Fig. 5.16).

To extend the flat pattern activity further, students were asked to select the same rectangles they had used for their first cage to see if they could tape the walls together in a different way. Students were encouraged to lay out the pieces and imagine folding up the sides around their creature. All students found at least one alternative arrangement. Eli experimented with several arrangements for DJ Woof (see Fig. 5.17). He said he imagined "rolling it up like a present" and then putting up the sides. He also noticed that as long as he had four walls "lined up", he could roll up his creature,

Fig. 5.16 Mr. Plow in his cage

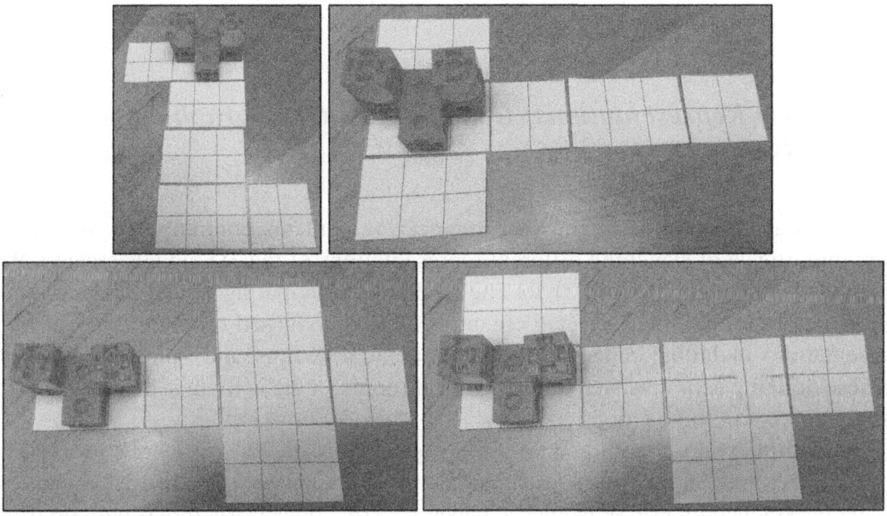

Fig. 5.17 Cages for DJ Woof

and the "wings" or the remaining two sides could be attached anywhere. "They did not even have to be opposite each other".

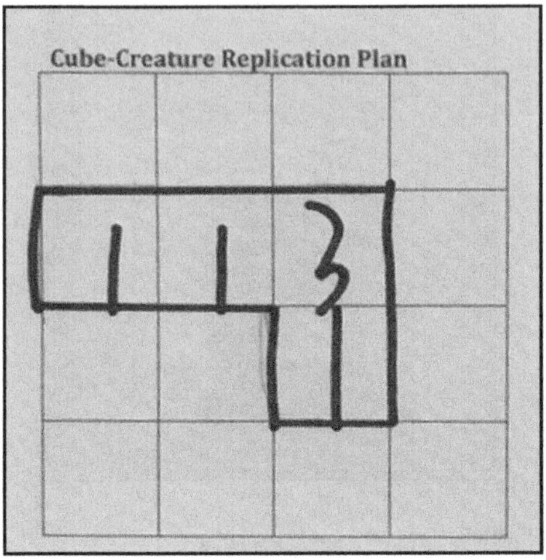

Fig. 5.18 Coded plan for Izzy

5.4.4 2D Representations: Coded Plans, Orthographic and Isometric Sketches

The spatial intervention studies for engineering students continued to emphasize the importance of sketching different representations of the 3D models created. We provided several opportunities for the students participating in the cube-creature activities to sketch their selected creatures and others based on three types of representations including coded plans, orthographic drawings, and isometric sketches. These drawing activities were interspersed throughout the activities, but are grouped together here in this section.

Coded plans are a common way for engineering students to learn to represent and define simple structures constructed by blocks (Lieu & Sorby, 2015). The plans, usually sketched on grid or isometric paper, describe how a structure is built up from its base. They are created by outlining the base of the structure and then the height of blocks is recorded numerically at each location. Both Sack (2013) and Patkin (2013) describe the use of coded plans or "top-view numeric coding" of cube structures with elementary students. Here, we used coded plans as a way for students to begin creating identification and replication plans for their selected cube-creature.

The students we worked with were asked to create a 2D coded plan which could be used to "manufacture" their creature. Through discussion and examples, the students traced the outline of their creatures' "footprint" and recorded whether the height was 1, 2, or 3 cubes tall in each square (see Fig. 5.18). To confirm their instructions, students traded plans with partners, manufactured a creature by building it up from the plan, and compared it to the original.

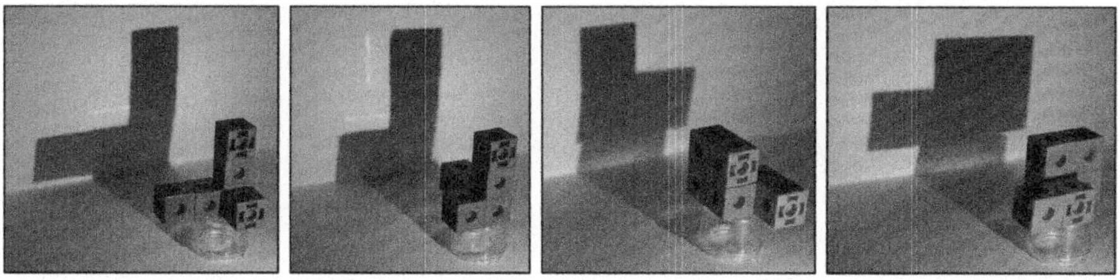

Fig. 5.19 Shadow creatures to see the front and side views

The second form of 2D representation came in the form of orthographic projection drawings (top–front–side). Such drawings are common in engineering graphics and are used to represent 3D structures on 2D surfaces. Each of the three views shows two dimensions. That is, the top view shows width and depth, the front view shows width and height, and the side view shows depth and height. Together, the three orthographic views require visualizing a single object from different perspectives. Although we anticipated that integrating the multiple perspectives would be challenging for primary-aged students, we attempted to mitigate difficulties through the original cube-creature constraints that eliminated the drawing issues associated with holes and overhangs. Spatial studies with elementary-aged children have used orthographic views with success (Moss et al., 2016; Sack, 2013). The multiview was connected to the students' experiences of tracing objects (e.g. hands, shapes, stencils) and making shadow puppets with their hands and objects around the room (e.g. coffee mug). The students predicted and described what the outlines of the projected images might look from different perspectives (e.g. the mug handle was not always visible). Using a similar approach, the students predicted, tested, and then drew what their creature would look like when shining a light on the "face" (i.e. front view) and "profile" (i.e. side view). Note: The top view was already created by tracing the base on the coded plans. Projections of Izzy and Squeegy are shown in Fig. 5.19. The set of coded plans and orthographic projection drawings are shown in Fig. 5.20.

The final form of 2D representation was for students to sketch perspective drawings of their creatures on isometric paper. Although this form of drawing is common in engineering, research with children reveal that such drawings pose difficulties for primary school children (Gutiérrez, 1996). Students were given guided assistance for using isometric paper starting with simple drawings of one cube, two cubes side by side, two stacked cubes, and so on. The students became progressively better at orienting the 3D structures to view them from the front edge, and then sketching the structures by following the outline. We noted a great range of motivation in drawing with isometric paper. While all students struggled initially, only a few persevered through multiple attempts to create a successful sketch (see Fig. 5.21).

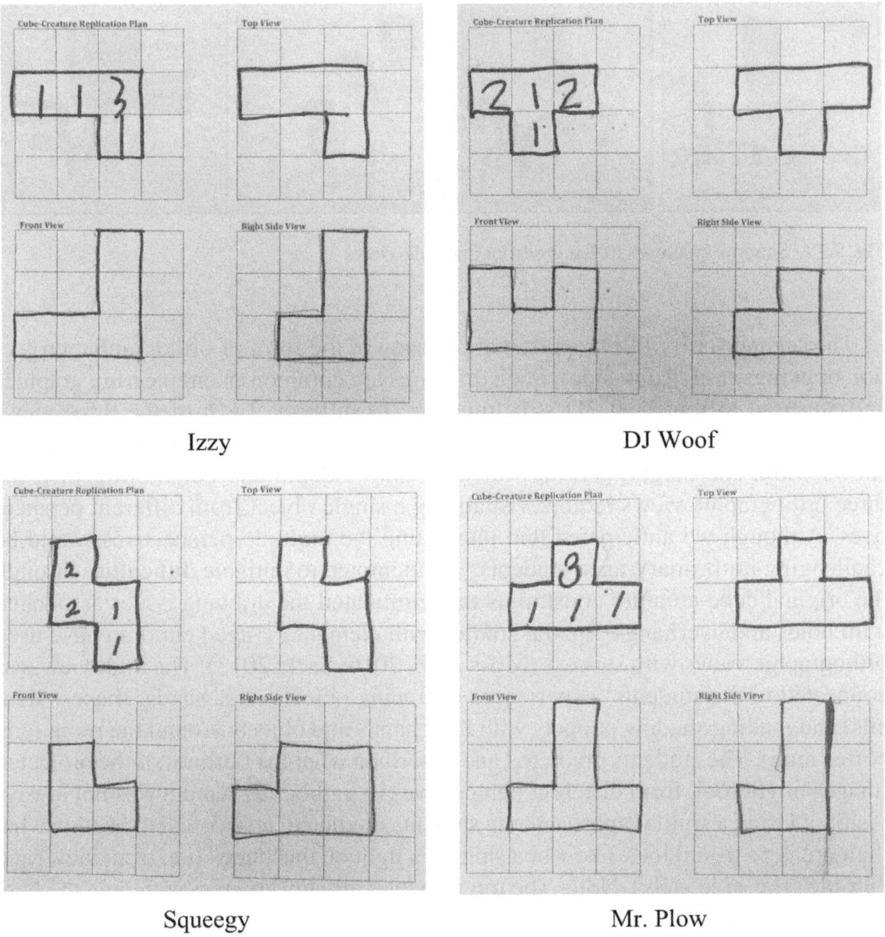

Fig. 5.20 Creature identification plans

In the previous section, we illustrated our spatial skills framework for young engineers through a series of spatial activities. In developing the activities to illustrate the framework, we used available literature, and relevant tasks created in field-tested and research-based items. In particular, we noted the importance of modelling 3D objects and using them for activities related to rigid transformations including rotation and symmetry, dynamic transformations through paper folding, and with many opportunities to sketch 3D structures from multiple perspectives and in multiple forms including coded plans, orthographic views, and perspective drawings.

Fig. 5.21 Jonas's drawings of Squeegy on isometric paper

5.5 Concluding Remarks

The early childhood classroom has an abundance of opportunities for children to develop spatial skills while using a variety of materials such as unit blocks, linking cubes, Lego, K-Nex. In many instances, these materials are accessed by children as part of open-ended and free-play activities. While such activity is important, our analysis of spatial assessment measures and spatial intervention programs for engineering students suggested that spatial skills that are not necessarily developed during free play. Given that visualization is one of the recognized engineering habits of mind, it is imperative that young children's potential for spatial skill development is fully realized.

We analyzed the spatial skills most frequently assessed at the post-secondary level and provided child-friendly measures of the same or similar skills. In particular, our analysis suggested four key areas for spatial skill development of current (and future) engineers. First, 3D modelling should involve building with imposed constraints or in response to problems presented or posed. Second, the construction of the 3D models should be used as the basis for description and comparison while employing physical and mental rotation, and comparing forms of symmetry. Third, spatial skill development should also include dynamic transformation such as paper folding and cross- sectioning. These skills require a flexibility of thinking to move between 2D and 3D structures. Finally, multiple perspective taking activities through different forms of drawing and 3D to 2D representational activities fulfils an essential aspect of spatial skill development. We believe that the information provided in this chapter can spark new assessment and intervention studies to help young children develop spatial skills needed to promote success in schooling and beyond.

Appendix A: Cube-Creature Spatial Reasoning Project

Introduction:
In this inquiry-based project students will be building, comparing, and representing Cube-Creatures in a series of activities.

Cube-Creature Composition:

Using a set of linking cubes, each student is asked to build at least three Cube-Creatures based on the following constraints:

- Creatures are only one colour
- Creatures consist of exactly 6 cubes
- Creatures stand on four cube-feet. No other body part touches the surface.
- Creatures stand on their own and stay connected when picked up.
- Creatures have no overhanging cubes.

Cube-Creature Classification:

In small groups, students pool their Cube-Creatures together and address the following questions and tasks posed by the teacher:

- Do all of the Cube-Creatures comply with the constraints?
- Are there any identical Cube-Creatures? Remove any duplicates (even if they are a different colour).
- How are the Cube-Creatures the same and how they are different? What are the ways you could sort and classify the Cube-Creatures?
- Agree on a way to classify and sort your Cube-Creatures into groups. Create missing Cube-Creatures to help extend your sorting groups.

Cube-Creature Identification and Replication Plans:

- Put a yellow sticker (eyeball) on the front of your creature and a blue sticker on one of its feet on the bottom.
- On the Cube-Creature planning sheet, write the name of your creature.
- Create a replication plan for your Cube-Creature (top left grid).

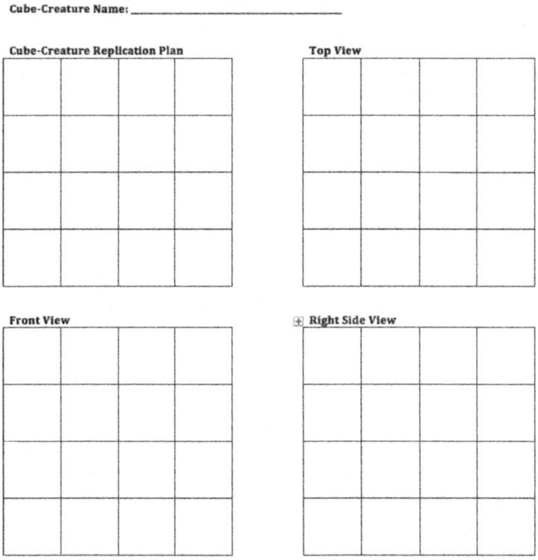

- Draw the front view, top view, and right side view of your Cube-Creature on the other grids.
- Is it possible to replicate a Cube-Creature from the top-front-side view? Do these three views provide enough information so that only one creature can be built?

Cube-Creature Cage Building:

Students create rectangular prism cages to keep their creatures and the public safe.

- Given pre-cut cage sides, create a 2D plan for a cage that snuggly fits their Cube-Creature. Start by placing their Cube-Creature on a 'floor' and then build the walls and roof.
- Will any other creatures fit into their cage? What are the characteristics of the creatures that will fit?
- Is there another way to construct the cage by attaching the walls in a different arrangement?

Cube-Creature 3D Drawings:

Students are given an opportunity to draw multiple views of their Cube-Creature using isometric grid paper.

- Learn to draw a cube, two cubes, stacked cubes on isometric paper.
- Draw the first layer (base) of their Cube-Creature.
- Draw their Cube-Creature from two different perspectives.

References

Bagiati, A., & Evangelou, D. (2016). Practicing engineering while building with blocks: Identifying engineering thinking. *European Early Childhood Education Research Journal, 24*(1), 67–85.

Bennett, G. K., Seashore, H. G., & Wesman, A. G. (1947). *Differential aptitude tests*. San Antonio, TX: Psychological Corporation.

Branoff, T. J., Hartman, N. W., & Wiebe, E. N. (2002). Constraint-based, three-dimensional solid modeling in an introductory engineering graphics course: Re-examining the curriculum. *Engineering Design Journal, 66*(1), 5–10.

Bruce, C. D., & Hawes, Z. (2015). The role of 2D and 3D mental rotation in mathematics for young children: What is it? Why does it matter? And what can we do about it? *ZDM Mathematics Education, 47*(3), 331–343. https://doi.org/10.1007/s11858-014-0637-4.

Casey, B., Andrews, N., Schindler, H., Kersh, J. E., Samper, A., & Copley, J. (2008). The development of spatial skills through interventions involving block building activities. *Cognition and Instruction, 26*(3), 269–309. https://doi.org/10.1080/07370000802177177.

Cheng, Y.-L., & Mix, K. S. (2014). Spatial training improves children's mathematics ability. *Journal of Cognition and Development, 15*(1), 2–11. https://doi.org/10.1080/15248372.2012.725186.

College Entrance Examination Board. (1939). *CEEB special aptitude test in spatial relations*. New York: Author.

Delgado, A. R., & Prieto, G. (2004). Cognitive mediators and sex related differences in mathematics. *Intelligence, 32*(1), 25–32.

Ekstrom, R. B., French, J. W., & Harman, H. H. (1976). *Manual for kit of factor referenced cognitive tests*. Princeton, NJ: Educational Testing Service.

Feng, J., Spence, I., & Pratt, J. (2007). Playing an action video game reduces gender differences in spatial cognition. *Psychological Science, 18*(10), 850–855.

Ferguson, C., Ball, A., McDaniel, W., & Anderson, R. (2008, November). *A comparison of instructional methods for improving the spatial-visualization ability of freshman technology seminar students*. Paper Presented at the 2008 IAJC-IJME International Conference, Nashville, TN. Retrieved from http://ijme.us/cd_08/PDF/37_IT305.pdf.

Field, B. W. (2007). Visualization, intuition, and mathematics metrics as predictors of undergraduate engineering design performance. *Journal of Mechanical Design, 129*(7), 735–743.

Frick, A., & Newcombe, N. S. (2015). Young children's perception of diagrammatic representations. *Spatial Cognition & Computation, 15*(4), 227–245. https://doi.org/10.1080/13875868.2015.1046988.

Frick, A., Ferrara, K., & Newcombe, N. S. (2013). Using a touch screen paradigm to assess the development of mental rotation between 3- and 5- years of age. *Cognitive Processing, 14*(2), 117–127. https://doi.org/10.1007/s10339-012-0534-0.

Gorska, R., & Sorby, S. (2008, June). *Testing instruments for the assessment of 3-D spatial skills.* Paper Presented at the ASEE Annual Conference, Pittsburgh, PA.

Guay, R. B. (1977). *Purdue spatial visualization test: Rotations*. West Lafayette, IN: Purdue Research Foundation.

Gutiérrez, A. (1996). Children's ability for using different plane representations of space figures. In A. R. Batturo (Ed.), *New directions in geometry education* (pp. 33–42). Brisbane, Australia: Centre for Math. and Sc. Education, Q. U. T.

Harris, J., Hirsh-Pasek, K., & Newcombe, N. S. (2013a). Understanding spatial transformations: Similarities and differences between mental rotation and mental folding. *Cognitive Processes, 14,*(2) 105–115.

Harris, J., Newcombe, N. S., & Hirsh-Pasek, K. (2013b). A new twist on studying the development of dynamic spatial transformations: Mental paper folding in young children. *Mind, Brain and Education, 7*(1), 49–55.

Hawes, Z., LeFevre, J., Xu, C., & Bruce, C. D. (2015a). Mental rotation with tangible three-dimensional objects: A new measure sensitive to developmental differences in 4- to 8-year-old children. *Mind, Brain, and Education, 9*(1), 10–18.

Hawes, Z., Moss, J., Caswell, B., & Poliszczuk, D. (2015b). Effects of mental rotation training on children's spatial and mathematics performance: A randomized controlled study. *Trends in Neuroscience and Education, 4,* 60–68. https://doi.org/10.1016/j.tine.2015.05.001i.

Hegarty, M., Kriz, S., & Cate, C. (2003). The roles of mental animations and external animations in understanding mechanical systems. *Cognition & Instruction, 21*(4), 325–360.

Hoyek, N., Collet, C., Fargier, P., & Guillot, A. (2012). The use of the Vandenberg and Kuse Mental Rotation Test in children. *Journal of Individual Differences, 33*(1), 62–67.

Hsi, S., Linn, M. C., & Bell, J. E. (1997). The role of spatial reasoning in engineering and the design of spatial instruction. *Journal of Engineering Education, 86*(2), 151–158.

Hungwe, K. N., Sorby, S., Durmmer, T., & Molzon, R. (2007). Preparing K-12 students for engineering studies by improving 3-D Spatial Skills. *The International Journal of Learning, 14*(2), 127–135.

Kersh, J., Casey, B. M., & Mercer Young, J. (2008). Research on spatial skills and block building in boys and girls: The relationship to later mathematics learning. In O. N. Saracho & B. Spodek (Eds.), *Contemporary perspectives on mathematics in early childhood* (pp. 233–252). Charlotte, NC: Information Age Publishing.

Koch, D., & Sanders, M. (2011). The effects of solid modeling and visualization on technical problem solving. *Journal of Technology Education, 22*(2), 3–21.

Levine, S., Ping, R., Young, C., & Ratliff, K. (unpublished manuscript). Toward a mental folding assessment.

Lieu, D. K., & Sorby, S. A. (2015). *Visualization, modeling, and graphics for engineering design* (2nd ed.). Boston, MA: Cengage Learning.

Lord, T. R., & Garrison, J. (1998). Comparing spatial abilities of collegiate athletes in different sports. *Perceptual and Motor Skills, 86,* 1016–1018.

Lucas, B., & Hanson, J. (2016). Thinking like an engineer: Using engineering habits of mind and signature pedagogies to redesign Engineering Education. *International Journal of Engineering Pedagogy, 6*(2), 4–13.

Marunić, G., & Glažar, V. (2014). Improvement and assessment of spatial ability in engineering education. *Engineering Review, 34*(2), 139–150.

Mohler, J. L. (2008). A review of spatial ability research. *Engineering Design Graphics Journal, 72*(3), 19–30.

Moss, J., Bruce, C. D., Caswell, B., Flynn, T., & Hawes, Z. (2016). *Taking shape: Activities to develop geometric and spatial thinking Grades K-2*. Don Mills, ON: Pearson.

Miller, C. L. (1996). A historical review of applied and theoretical spatial visualization publications in engineering graphics. *Engineering Design Graphics Journal, 60*(3), 12–33.

Mix, K., & Cheng, Y. (2012). The relation between space and math: Developmental and educational implications. In J. B. Benson (Ed.), *Advances in child development and behaviour* (Vol. 42, pp. 197–243). Burlington: Academic Press, Elsevier Inc.

National Research Council. (2006). *Learning to think spatially*. Washington, DC: National Academies Press.

Newcombe, N. S. (2010). Picture this: Increasing math and science learning by improving spatial thinking. *American Educator, 34*(2), 29–35, 43.

Newcombe, N. S., & Frick, A. (2010). Early education for spatial intelligence: Why, what, and how. *Mind, Brain, and Education, 4*(3), 102–111.

Newcombe, N. S., & Shipley, T. F. (2015). Thinking about spatial thinking: New typology, new assessments. In J. S. Gero (Ed.), *Studying visual and spatial reasoning for design creativity* (pp. 179–192). New York, NY: Springer.

Newcombe, N. S., Uttal, D. H., & Sauter, M. (2013). Spatial development. In P. D. Zelazo (Ed.), *Oxford handbook of developmental psychology, Vol. 1: Body and mind* (pp. 564–590). New York: Oxford University Press.

Onyancha, R. M., Derov, M., & Kinsey, B. L. (2009). Improvements in spatial ability as a result of targeted training and computer-aided design software use: Analyses of object geometries and rotation types. *Journal of Engineering Education, 98*(2), 157–167.

Patkin, D. & Fadalon, L. (2013). Developing third grade boys and girls' spatial ability by means of an extra-curricular teaching unit. *Journal of the Korean Society of Mathematical Education Series D: Research in Mathematical Education, 17*(2), 99–118.

Ratliff, K. R., McGinnis, C. R., & Levine, S. C. (2010). *The development and assessment of cross sectioning ability in young children*. Paper Presented at the 32nd Annual Conference of the Cognitive Science Society. Portland, Oregon.

Sack, J. J. (2013). Development of a top-view numeric coding teaching-learning trajectory within an elementary grades 3-D visualization design research project. *The Journal of Mathematical Behavior, 32*(2), 183–196.

Sanchez, C. A., & Wiley, J. (2014). The role of dynamic spatial ability in geoscience text comprehension. *Learning and Instruction, 31,* 33–45.

Sarama, J., & Clements, D. H. (2009). *Early childhood mathematics education research: Learning trajectories for young children*. New York: Routledge.

Sinclair, N., & Bruce, C. D. (2015). New opportunities in geometry education at the primary school. *ZDM Mathematics Education, 47*(3), 319–329.

Sorby, S. A. & Baartmans, B. J. (1996). A course for the development of 3-D spatial visualization skills. *Engineering Design Graphics Journal, 60*(1), 13–20.

Sorby, S. A. (2009). Educational research in developing 3-D spatial skills for engineering students. *International Journal of Science Education, 31*(3), 459–480.

Sorby, S. A., & Baartmans, B. (2000). The development and assessment of a course for enhancing the 3-D spatial visualization skills of first year engineering students. *Journal of Engineering Education, 89*(3), 301–307.

Sorby, S. A., & Veurink, N. (2010). Are the visualization skills of first-year engineering students changing? In *Proceedings of the 117th ASEE Conference and Exposition.*

Sorby, S., Casey, B., Veurink, N., & Dulaney, A. (2013). The role of spatial training in improving spatial and calculus performance in engineering students. *Learning and Individual Differences, 26,* 20–29.

Taylor, H. A., & Hutton, A. (2013). Think3d!: Training spatial thinking fundamental to STEM education. *Cognition and Instruction, 31*(4), 434–455. https://doi.org/10.1080/07370008.2013.828727.

Tsutsumi, E. (2005). Evaluation of students' spatial abilities using mental cutting test. *International Journal of Technology in Engineering Education, 2*(2), 77–82.

Tzuriel, D., & Egozi, G. (2010). Gender differences in spatial ability of young children: The effects of training and processing strategies. *Child Development, 81*(5), 1417–1430.

Uttal, D. H., Meadow, N. G., Tipton, E., Hand, L. L., Alden, A. R., Warren, C., & Newcombe, N. S., (2013). The malleability of spatial skills: A meta-analysis of training studies. *Psychological Bulletin, 139*(2), 352–402.

Vandenberg, S. G., & Kuse, A. R. (1978). Mental rotations, a group test of three-dimensional spatial visualization. *Perceptual and Motor Skills, 47*(2), 599–604.

Verdine, B. N., Golinkoff, R. M., Hirsh-Pasek, K., & Newcombe, N. S. (2017). Spatial skills, their development, and their links to mathematics. *Monographs of the Society for Research in Child Development, 82*(1), 7–30.

Wai, J., Lubinski, D., & Benbow, C. P. (2009). Spatial ability for STEM domains: Aligning over 50 years of cumulative psychological knowledge solidifies its importance. *Journal of Educational Psychology, 101,* 817–835.

Webb, R. M., Lubinski, D., & Benbow, C. P. (2007). Spatial ability: A neglected dimension in talent searches for intellectually precocious youth. *Journal of Educational Psychology, 99*(2), 397–420.

Wei, W., Yuan, H., Chen, C., & Zhou, X. (2012). Cognitive correlates of performance in advanced mathematics. *British Journal of Educational Psychology, 82,* 157–181.

Wulf, W. A. (1999). The image of engineering. *Issues in Science and Technology, 15*(2). Retrieved from http://issues.org/15-2/wulf-2/.

Chapter 6
Identifying Engineering in a PreK Classroom: An Observation Protocol to Support Guided Project-Based Instruction

Aikaterini Bagiati and Demetra Evangelou

Abstract This chapter presents an early engineering curriculum for the PreK classroom, justifies the developmental appropriateness of the curriculum by presenting relevant research studies, and concludes with the introduction of an observation protocol to be used by class teachers to identify and evaluate engineering learning. In early education engineering related resources are still very limited. Scattered activities or small scale engineering lesson plans can be found for a teacher to use in class mostly lacking appropriate assessment tools. Additional obstacles center on teacher preparedness and ensuing "discomfort" with engineering content, terminology, and procedures. Limited exposure to engineering content reputed as a difficult discipline requiring rigorous specialization, makes most teachers apprehensive and very reluctant to explore and introduce it in the curriculum. The early engineering curriculum discussed here was developed and implemented in a PreK classroom for 4 months. Student learning and the teacher experience were at the center of the research. The proposed observation protocol was designed in alignment with the research findings. Observation is a powerful tool and in this case it is used to inform assessment in early education. The protocol is expected to assist PreK teachers in developing deep understanding of how to identify and evaluate engineering learning in class.

A. Bagiati (✉)
Massachusetts Institute of Technology, Cambridge, MA, USA
e-mail: abagiati@mit.edu

D. Evangelou
Democritus University of Thrace, Alexandroupolis, Greece
e-mail: evangel@psed.duth.gr

L. English and T. Moore (eds.), *Early Engineering Learning*, Early Mathematics Learning and Development, https://doi.org/10.1007/978-981-10-8621-2_6

6.1 Introduction

Recent attempts to reform PreK-12 engineering education in the US, as well as in other European and Asian countries, start at the early ages, even though resources are still very limited (Bagiati et al., 2015; DeJarnette, 2012; NGSS, 2013). Some resources have already been developed for the early education teachers to use; they are, however, very rarely accompanied by appropriate assessment tools (Bagiati, Yoon, Evangelou, & Ngambeki, 2010). In the case of the US, additional obstacles center on teacher preparedness and ensuing "discomfort" with engineering content, terminology, and procedures (Bagiati, 2011; Bagiati & Evangelou, 2015; Brophy, Klein, Portsmore, & Rogers, 2008). Limited exposure to engineering content, reputed as a difficult discipline requiring rigorous specialization, makes most teachers apprehensive and reluctant to explore and introduce it in the curriculum (Bagiati, 2011; Bagiati et al., 2010; Hsu, Purzer, & Cardella, 2011).

Bringing Engineering in a PreK class appears to require work on two different levels; (a) age appropriate curriculum development, and (b) teacher training and support (Culver, 2012; Duncan, Diefes-Dux, & Gentry, 2011). This chapter introduces an early engineering curriculum called "Puppeteering to Engineering" (P2E), examines the developmental appropriateness of the curriculum, and presents a Pre-Kindergarten Engineering Observation Protocol (PREEOP) to assist early education teachers identify engineering related learning. P2E is an early engineering curriculum based on three research studies, which was implemented in a PreK classroom for 4 months with a strong focus on the student learning and the teacher experience. Observation is a powerful tool and in this case, it is designed to inform assessment in early education. The protocol was designed in alignment with the research findings that emerged from the curriculum implementation in class, and it is expected to assist PreK teachers in developing a deeper understanding of how to identify and evaluate engineering learning in class.

6.2 Preliminary Studies

The foundational studies that led to the development of the P2E curriculum began in 2007, when a review of the literature suggested that "developmental theory and empirical research firmly support the assumption that objects and their use by children constitute a universal part of development and learning" (Brophy & Evangelou, 2007). Considering the nascent stage of early engineering at the time, we began by attempting to establish whether bringing engineering related content in a PreK classroom was developmentally appropriate, and what it would take to bring such content in early childhood classes.

6.2.1 Study 1: Identifying Precursors to Engineering Thinking

According to the Accreditation Board for Engineering and Technology "Engineering design is the process of devising a system, component, or process to meet desired needs. It is a decision-making process (often iterative), in which the basic science and mathematics and engineering sciences are applied to convert resources optimally to meet a stated objective" (ABET, 2016). Although not all engineers, or teams of engineers, follow the exact same steps when designing, there are some steps that are considered fundamental in this process. A typical cycle of this process usually starts by a given problem, or an initial thought for a new product or system. Then the engineer would do some brainstorming of new ideas on how to solve the problem, and proceed with identifying related preexisting work done on the same field. Building a "model" or a "prototype" would follow, and then the engineer would test the model, consult other resources for improvement, gather more information, rebuild and retest, until satisfied with the final product. Figure 6.1 describes such a Fundamental Design Process.

Block building time in PreK was the setting for the first study. Blocks have been considered a staple to early childhood education for decades now, and they

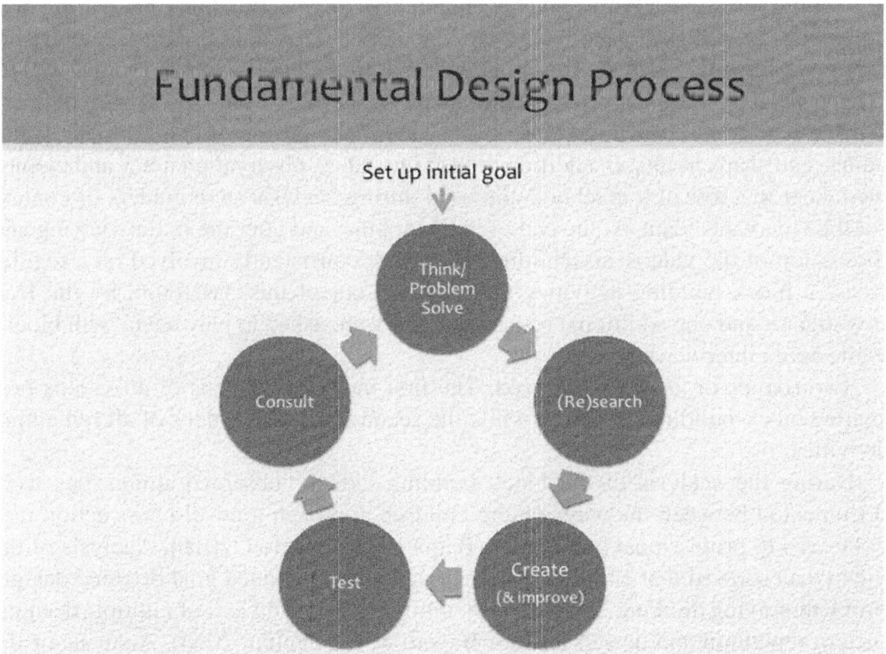

Fig. 6.1 Fundamental design process common across engineering disciplines (Bagiati & Evangelou, 2016)

are already considered a great learning tool for children (Tepylo, Moss, & Stephenson, 2015). Among other skills, block building enhances communication, problem solving, development of rationale, development of math and science concepts, and construction skills (Bagiati & Evangelou, 2016; Newburger & Vaughan, 2006). Synthesizing all the aforementioned domains can lead to the development of certain process models regarding the way children's block constructions could be designed, built and improved (Bagiati & Evangelou, 2016). The Design Process model constitutes one of the core concepts of engineering; so blocks, very commonly present in PreK classrooms, seem to be "one of the best tools to use in order to work towards the development of such a model." (Bagiati & Evangelou, 2016).

The goal of this first study was to examine whether young children presented instances of precursors to engineering behavior while playing, as seen through similarities to the Fundamental Design Process, as presented in Fig. 6.1. More specifically this study examines children's free play, especially building with blocks, to identify similarities between the way children build and the Fundamental Design Process. "Children's play naturally employs skills of observation and experimentation that lead to the development of intuitive models for how things work" (Brophy & Evangelou, 2007), therefore, while observing the children, authors also tried to identify patterns of engineering related behaviors and procedural work models that may indicate precursors to engineering thinking.

In this qualitative observational study of free play, the focus was on informal play as a setting for active learning and development of engineering thinking. During data collection, 18 children, aged 3–5, were videotaped daily during their free playtime for approximately 2 h in the morning and 1.5 h in the afternoon for 4 months. Data were collected through a series of naturalistic field observations and the use of open, semi structured and structured material such as blocks, puzzles, Lego™ blocks, water tables, and snap circuits. In addition to the videotapes, observation notes and memos were kept in a research notebook and used during analysis as reminders of context for the videotaped data. At the end of the 4 months, and after the initial viewing and discussion of the videos, six children who were consistently involved on a regular basis in block building activities were invited out of the classroom, by the lead investigator and one additional researcher, and were asked to play again with blocks while being interviewed.

Two rounds of analysis followed. The first included analysis of all videos presenting block-building activities, while the second included videos of all remaining activities.

During the analysis of the block building videos "observed similarities were documented between the ways young children approach a novel construction task compared to professional engineers" (Bagiati & Evangelou, 2016). Analysis of the video data showed that children demonstrated and articulated goal oriented design, problem-solving thinking, innovation stemming out of synthesis of multiple designs, pattern repetition and design testing (Bagiati & Evangelou, 2016). Analysis of the videos presenting children engaged with open (sensorial) materials, semi-structured play and structured play, also revealed behaviors related to engineering thinking. Behaviors observed were (a) asking questions and stating goals, (b) explaining how things are built/work, (c) constructing/making things, (d) solving problems, and (e) evaluating design (Bairaktarova, Evangelou, Bagiati, & Brophy, 2011). These results suggest that trained adults can capitalize on spontaneous play as opportunities for introducing engineering.

6.2.2 *Study 2: Identifying Appropriate Types of Resources for Use in Class*

Design, typically resulting in the creation of human–made artifacts, is at the core of engineering. The primary focus of our second study was understanding "engineering thinking as it is revealed in young children's activities and interactions with the world of artifacts" (Evangelou, Dobbs-Oates, Bagiati, Liang, & Choi, 2010). More specifically we aimed to identify the knowledge young children have about human-made artifacts, understand how this knowledge develops, and understand if the development gets better supported by the interaction with 3D artifacts or 2D representations.

For the purpose of the study 35 children ages 4–5 from 6 different PreK classrooms were interviewed to identify their prior knowledge of human-made artifacts. The children were randomly assigned to three different conditions. "Each condition included the same set of 13 different artifacts that were either artistically rendered in black ink on white paper (sketch condition), included in a children's storybook (book condition), or had the real artifact itself (tangible object condition). Children's exploration and interactions were videotaped and analyzed to see which, if any, of the three conditions would appear to stimulate and encourage early engineering thinking the most" (Evangelou et al., 2010). Initial hypothesis was that interaction with the tangible objects would lead to more exploration time and to more explanations from the children in regard to the function and use of the artifact. "Findings showed that this condition elicited the longest discussions and interactions with the artifacts, and it was also the condition during which children were demonstrating more knowledge and ideas with regard to possible functions of the artifacts. Regarding whether there was a condition that stimulated more interest toward specific artifacts, no clear pattern among the three conditions appeared" (Evangelou et al., 2010).

Findings of this study are well aligned with the well-documented benefits of exploratory play in early education (Bonawitz et al., 2009; Cook, Goodman, & Schulz, 2011; Jennings, Harmon, Morgan, Gaiter, & Yarrow, 1979; Piaget, 1929), and support the idea of using artifacts, especially in the form of tangible-objects, as developmentally significant in promoting early engineering thinking through exploration.

6.2.3 Study 3: Identifying Open Early Engineering Resources

The ongoing global debate about K-12 engineering education gives rise to questions about appropriate educational materials (DiFransesca, Lee, & McIntyre, 2014; Jahan & DeJarnette, 2014; Katehi, Pearson, & Feder, 2009). "Introducing engineering in the early years entails recognition of the need for teachers to understand its content and poses the challenge of preparing teachers to incorporate engineering education into their practice" (Bagiati et al., 2010) and directly affects future teacher professional development. Seeking information in books, journals, magazines, and following professional development programs offered by universities, school districts, and other educational entities, seems to have largely been replaced by Internet inquiries for novel materials (Hedtke, Kahlert, & Schwier, 2001; Recker, 2006; Bagiati, Yoon, Evangelou, & Ngambeki, 2010; Bagiati, Yoon, Evangelou, Magana, et al., 2015). In 2009 we undertook a systematic study of open resources developed by entities formally related to education that were becoming available to early education teachers, We initially sought resources in English and followed with Chinese, French, Greek, Korean, Spanish and Turkish in 2010 (Bagiati et al., 2011). The search was repeated again in 2014 resulting in the close examination of resources in 7 different languages, namely Arabic, Chinese, English, French, Greek, Korean and Spanish.

The first search for open online resources for preschool through grade 12 (PreK-12) engineering materials, conducted in English in 2010, revealed a wide variety of Web sites and online documents that included curricula, lesson plans, and descriptions of activities. Narrowing the search to the PreK-3 level revealed that the pedagogically and content-reliable sources available are limited in number and may be difficult to identify among the plethora of information" (Bagiati et al., 2010). Websites were presented as containing PreK-12 engineering curricula, but upon closer examination of the PreK-3 content, indicated that most were portals leading to sites, which included scattered free-standing activities unrelated to each other or to complete lesson plans. By the end of 2010 no curricula for a PreK classroom were located, and the dearth of early engineering recourses highlighted the need for an early engineering curriculum even more, and led to the development and implementation the P2E curriculum in class by our group. In 2014 an additional curriculum for Kindergarten and two early elementary curricula were located (Fig. 6.2).

6.3 Puppeteering to Engineering: An Early Engineering Curriculum

Our preliminary studies led to the conclusion that introducing engineering at the PreK classroom appeared to be developmentally appropriate, and that a project including hands-on components and interaction with tangible artifacts would be the most beneficial approach. Analysis of existing resources supported the need for a cohesive

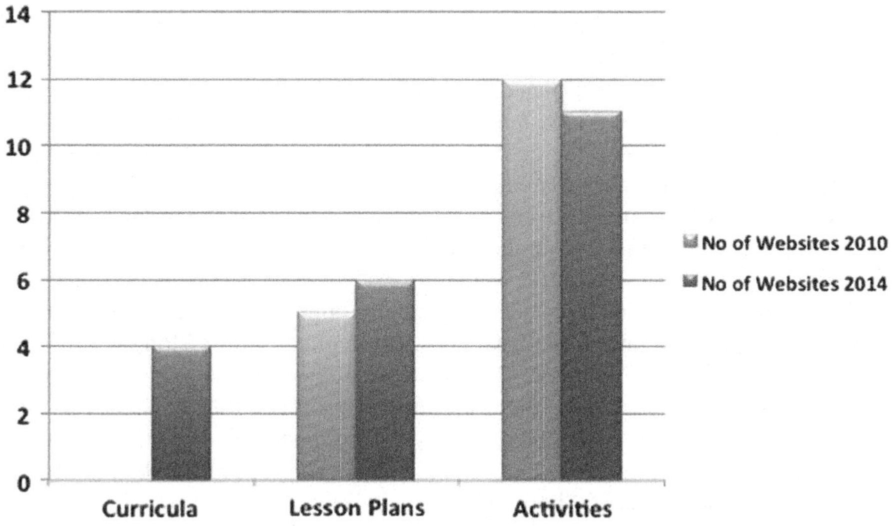

Fig. 6.2 Number of websites presenting early engineering curricula, lesson plans and activities in English as identified in 2010 and 2014

curriculum that could be delivered to the teacher accompanied by appropriate in-service training prior to implementation.

6.3.1 Curriculum Development

Puppeteering to Engineering (P2E) is designed to address STEM integration at early education, with primary emphasis on the engineering component. It employs a view of engineering "as a disciplinary domain that uses math, science, and technology as tools, but which also requires synthetic ability, design, problem solving, organization, and construction skills, and incorporates various types of communication as well. From this viewpoint, engineering is thus more than the sum of the STEM parts" (Bagiati, 2011).

P2E was developed based on two curricular frameworks, namely *The Creative Curriculum* (Dodge & Colker, 1996) and *The Project Approach* (Katz & Chard, 2000). *The Creative Curriculum* is a holistic teacher-driven framework (Dodge & Colker, 1996), widely used in the US, and also used in the classroom in which the research study took place. This framework would allow the research group to establish "a stimulating early engineering learning environment and to address development of selected pre-planned STEM knowledge and skills" (Bagiati & Evangelou, 2011). *The Project Approach* is a child-driven framework designed to complement other teacher-driven preschool educational curricular frameworks (Katz & Chard, 2000), and in P2E "it was employed in order to complement *The Creative Curricu-*

lum and to offer to the children a child-driven design project experience" (Bagiati & Evangelou, 2011).

P2E consists of 24 lesson plans presented to the children through two gender-neutral puppets, Sam and Andy; hence the name of the curriculum. The choice of having puppets introducing the activities has been very deliberate, as the multiple benefits of using puppets in early education are long known. Among others, puppets have been long used to make a topic more engaging, but also have been reported to be a narration method that encourages the children to express themselves more and in greater comfort (Korošec, 2012; Majaron, 2012).

The 24 lesson plans address developmentally appropriate science, technology, engineering and math concepts and practices, all integrated within a long–term design project. Many different thematic ideas were examined before deciding what the final project would be. Among them the most appropriate themes appeared to be "Lets build a city" and "Lets travel far away". The theme that we selected for this particular implementation was "Let's build a city". The idea of building a city was considered appropriate as the "building" and "city" concepts were already familiar in the PreK class, and the topic could be open-ended enough to allow children to discuss both the interiors of buildings of their choice as well as the environment (Bagiati, 2011). Furthermore, this theme was considered to be a good fit to our plans because the teacher in this particular classroom had not implemented a similar construction project in the last two years; therefore, the topic would be new to the children (Bagiati, 2011). To implement the project, Sam and Andy, the two puppets, introduced the following scenario to the children in class.

> Sam had some dolls and some cars and other toys in a dollhouse. Sam threw a birthday party and now there are more dolls, cars, and other toys, so the dollhouse is not big enough. Sam now has four dolls, three cars, one dog and one ball. Sam is discussing this situation with Andy, who is an engineer, about new ideas.

Every lesson plan in P2E addresses ten items (Bagiati, 2011).

1. New engineering related concepts and terminology to teach in class
2. Goals and Objectives
3. Required Materials
4. Setting
5. Step-by-step procedures
6. Plan for independent practice
7. Closure and Reflection
8. Assessment based on objectives
9. Possible connections to other subjects
10. Image(s) to be placed on a cardboard in class.

Appendix 1 presents the full set of lesson plans along with engineering related concepts to be introduced, and the engineering related rationale behind every concept, however all lesson plans can be found in full detail at (Bagiati, 2011) or can be downloaded from www.puppetengineering.com.

6.3.2 Classroom Implementation and Teacher Preparation

Implementation of the P2E lasted 12 weeks, from September to December, and the early engineering lessons took place twice per week. The teacher, with the use of two puppets, introduced the integrated STEM content to class during large group (LG) discussion time. The use of *The Creative Curriculum* framework was employed at this part of instruction. A small group (SG) creative or design activity followed the large group discussion time. During that time *The Project Approach* was used as a framework for instruction (Bagiati, 2011; Bagiati & Evangelou, 2015).

In regard to teacher preparation, the P2E developer was transferring two lesson plans to the teacher every week, and about one week before the lesson plans would be implemented in class. A brief and a long meeting between the teacher and the developer would follow the delivery of the lesson plans. The brief meetings would typically occur immediately before the class time and the goal was to address any last-minute questions the teacher might have, while the long meetings included debriefing about lessons taught during the week, clarifying the teacher's questions with regard to the new lesson plans, and planning for subsequent lessons (Bagiati & Evangelou, 2015). In addition to the teacher one substitute teacher and a student in training were always present in the classroom, but different persons filled these two roles at different times;"it was therefore the teacher's responsibility to educate them in regards to tasks and roles she wanted them to undertake in class" (Bagiati & Evangelou, 2015).

6.3.3 Assessment of P2E: Research on STEM Learning and Findings

As the P2E was implemented in class, a research study on the implementation of P2E took place in order for our research group to evaluate the early engineering related learning outcomes. For the purpose of this study learning in early education, as defined by Katz, is a synthesis of knowledge, skills, dispositions and feelings; as follows:

Participants were 11 children with parental permission (10 boys and one girl), their parents, and the classroom teacher who introduced P2E in class. The children were from a mostly middle socio-economic area. Data for this study consisted of the researcher field notes from class collected during and after the P2E sessions; documentation of the children's work; a teacher journal; one teacher exit interview; and letters from parents.

Throughout data collection, it was obvious that children could demonstrate engineering-related behaviours even without verbalizing, either because they were building or creating something alone or because some of the children were not English speakers. When talking, children were observed using language to explain something to the teacher or the researcher, to share an initial construction goal, to give their input regarding a solution, to express a complaint or state a problem, or to help or consult

KNOWLEDGE. In early childhood, knowledge consists of facts, concepts, ideas, vocabulary, stories, and many other aspects of children's culture. Children acquire such knowledge from someone's answers to their questions, explanations, descriptions, and accounts of events, as well as through active and constructive processes of making the best sense they can of their own direct observations
SKILLS. Skills are small units of action that occur in a relatively short period of time and are easily observed or inferred. Physical, social, verbal, counting, and drawing skills are among a few of the almost endless number of skills learned in the early years. Skills can be learned from direct instruction or imitated based on observation, and they are improved with guidance, practice, repetition, drill, and actual application or use
DISPOSITIONS. Dispositions can be thought of as habits of mind or tendencies to respond to certain situations in certain ways. Curiosity, friendliness or unfriendliness, bossiness, generosity, meanness, and creativity are examples of dispositions or sets of dispositions, rather than of skills or items of knowledge. Accordingly, it is useful to keep in mind the difference between having writing skills and having the disposition to be a writer, or having reading skills and having the disposition to be a reader
FEELINGS. Feelings are subjective emotional states. Some feelings are innate (e.g., fear), while others are learned. Among feelings that are learned are those of competence, confidence, belonging, and security. Feelings about school, teachers, learning, and other children are also learned in the early years. (Katz, 1999, p. 3)

with each other, "but sometimes they just looked around for answers to solve their design problems or just intervened on another child's construction in order to bring their ideas to the task" (Bagiati & Evangelou, 2016). In order to identify STEM learning through demonstration, researchers were focusing on children's body language, and implied use of blocks, toys, or other materials. Data were qualitatively analyzed using the open coding method.

At the end of the first round of data analysis, and according to Katz's definition of learning, the four learning categories, related to early engineering, as presented in the following list, were identified.

STEM learning categories

1. Knowledge	Math, Science, Technology, Artifacts, Functions, Buildings, Materials, Construction, Design mental models, Visual representations, Engineering Vocabulary, Engineering process, Engineering profession, Self-competence, Classroom awareness (Bagiati & Evangelou, 2011, p. 4)
2. Skills	Multimodal learning, Questions development, Communication, Reasoning, Observation, Problem-solving, Technology usage, Drawing, Collaboration, Construction, Synthesis, Innovation, Improvement, Test/compare, Give/follow instructions, Reproduction (Bagiati & Evangelou, 2011, p. 4)
3. Dispositions	Preference towards construction, collaboration, theme, and materials, Requests for documentation of own work (Bagiati & Evangelou, 2011, p. 5)
4. Feelings	Enthusiastic, Pleased by their own work, Pleased by their own ideas, Pleased by parent participation, Like resources, Engaged, Bored/distracted, Frustrated by their own failure, Frustrated by other's actions/choices, Dislike resources (Bagiati & Evangelou, 2011, p. 5)

Taking a look at all learning categories above, it become obvious that some of them particularly refer to STEM learning (e.g. knowledge regarding the engineering profession), while others represent broader learning goals well stated in the traditional literature regarding learning goals in early education (e.g. developing self-competence). To establish the appropriateness of P2E it was therefore essential for our group to monitor whether the P2E curriculum would address the traditional learning expectations in addition to the STEM related ones. The development of positive or negative feelings, although it might not appear as relevant to STEM learning as the other categories, was absolutely essential to monitor, as the feelings developed through interaction with a specific content or activity will probably affect the child's long-term desire to further engage with anything similar.

At this point is should be noted that STEM learning instances, as identified in class, may reveal learning associated with more than one subcategory, i.e. a child talking about how to use a brick indicates knowledge related to buildings, artifacts and materials on the same time; and it also demonstrates communication, reasoning and construction skills. Therefore identifying the overlap between all subcategories was the second level of data analysis. The content and learning appeared to have been approached through three different avenues. Discussions and activities either addressed *STEM directly* or stimulated further discussions and activities about *artifacts* and *buildings*. Participation in P2E framed these three approaches and formed a larger holistic framework for learning that was developed in relation to STEM as well as to the broader disciplinary goals to be achieved in early education. Examining the overlap among them, the first overlap identified was *in artifacts and technology*. In the current case, technology was considered a subset of the artifacts category. A larger overlap was identified among the categories of *technology, artifacts, buildings, and engineering* (Bagiati, 2011). Overlap among all different approaches is presented in Fig. 6.3, where the blue box represent direct inferences to STEM content, and the pink and green represent indirect inferences via interactions with artifacts or building.

Examining the findings regarding STEM-related knowledge, we can see that learning relevant to the engineering process, the construction phase, functionality, visual representation, and design became apparent both through the direct and indirect approaches towards the STEM content. Furthermore two subcategories, self-competence and classroom-awareness, were placed within the larger framework of the P2E curriculum as they address broader developmental goals of early childhood education. Figure 6.4 presents how the STEM-related knowledge subcategories were placed within the P2E framework.

In regards to STEM-related skills, some of them can be directly related to the engineering design process (Fig. 6.5), and their development could be initiated and enhanced through discussions and activities approached through direct STEM discussion, or interactions with artifacts and buildings (Fig. 6.6). The remaining skills, although considered to be essential to STEM and developed through the P2E framework, address broader more traditional developmental goals and are therefore listed separately.

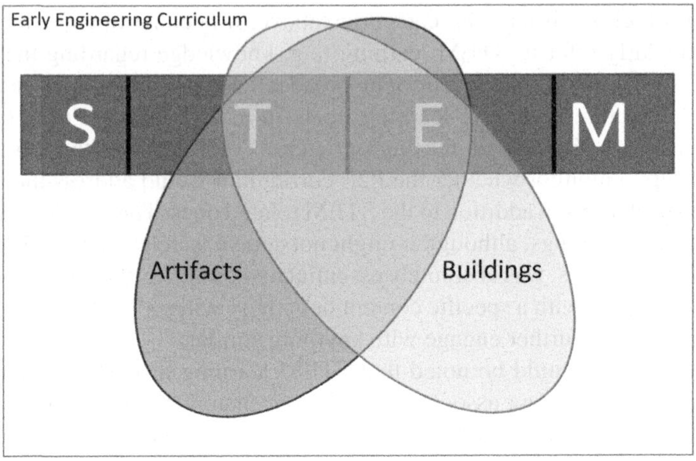

Fig. 6.3 A holistic early education framework through which STEM learning can be approached (Bagiati, 2011)

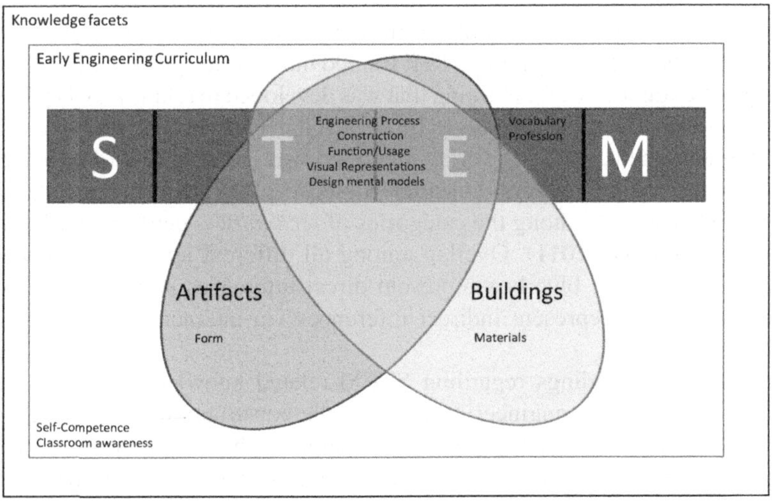

Fig. 6.4 STEM-related knowledge subcategories, as they appear to be developed within the P2E framework (Bagiati, 2011)

STEM-related dispositions, preferences towards construction activities, collaborations, working themes, and working materials, as well as STEM-related positive and negative feelings, emerged out of discussions and activities related to STEM, artifacts, or buildings as presented in Figs. 6.7 and 6.8. Children also demonstrated a desire to have their complete work documented many times throughout our data collection period in class.

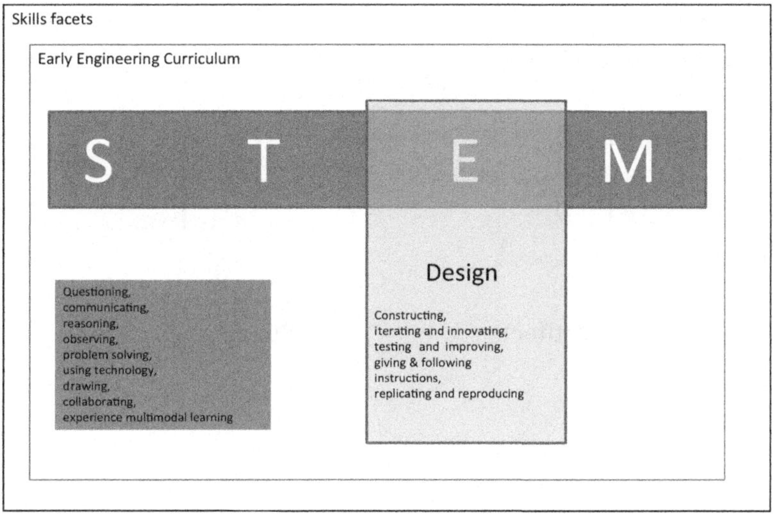

Fig. 6.5 Skills addressing broad developmental goals and skills related to the engineering design process (Bagiati, 2011)

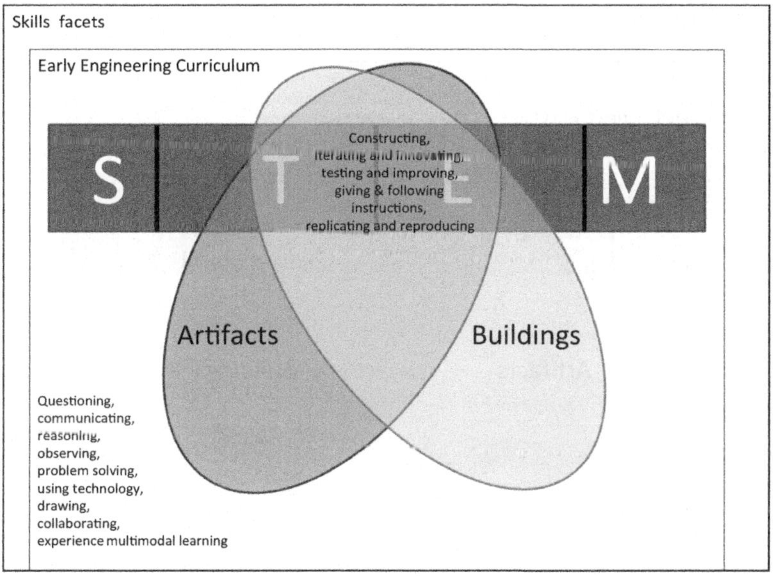

Fig. 6.6 STEM-related skills subcategories, as they appear to be developed within the P2E framework (Bagiati, 2011)

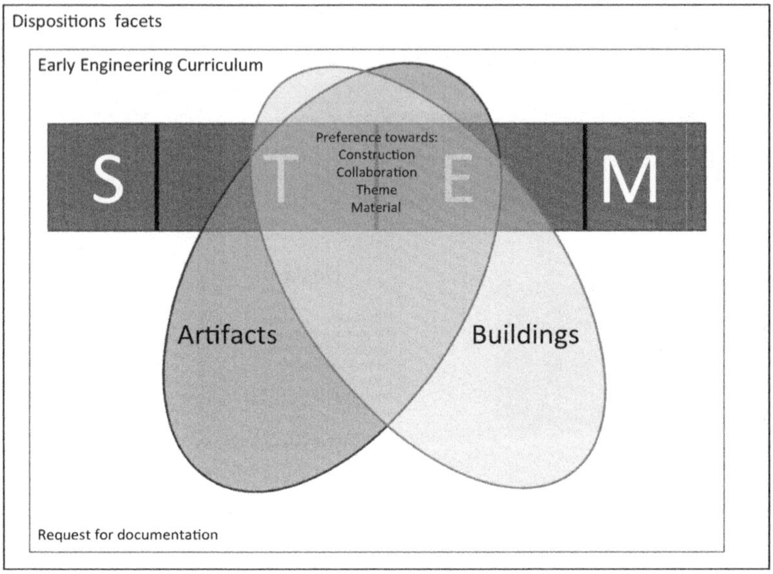

Fig. 6.7 STEM-related dispositions subcategories, as they appear to be developed within the P2E framework (Bagiati, 2011)

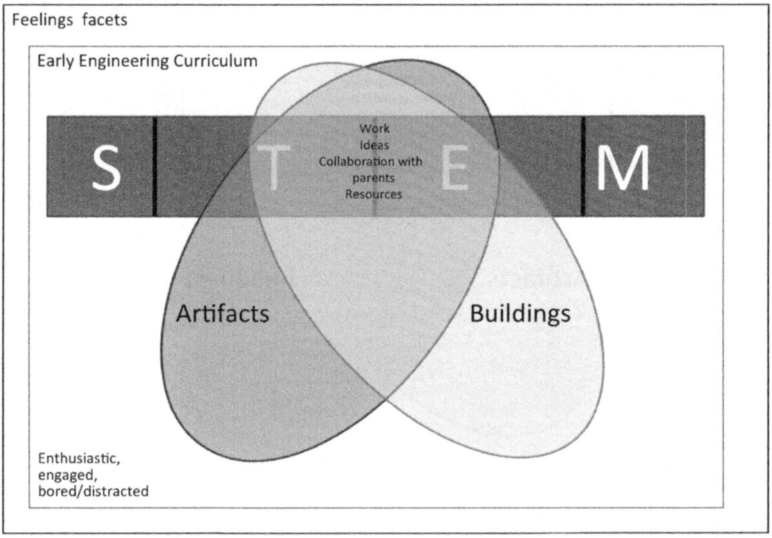

Fig. 6.8 STEM-related feelings subcategories, as they appear to be developed within the P2E framework Bagiati, 2011)

6.3.4 Developing the Pre-Kindergarten Engineering Observation Protocol (PREEOP)

Once all research studies had been completed, and after having spent more than 4 years in PreK classes observing and discussing early engineering activities with a large number of teachers, our group had now a much greater understanding of the STEM related learning that may take place in a PreK classroom. However it had also become very obvious to us that there is still a great disconnect between the STEM related learning that takes place in class, and the learning outcomes an early education teacher is formally trained to observe and identify. While early education teachers are extensively trained to monitor physical, social and emotional, cognitive, and language and literacy development (NAEYC, 2009), development in areas related to early engineering learning feels like being almost transparent. As a result we knew that the appropriate next step in order to properly facilitate the introduction of engineering in an early education classroom is to provide a new lens by developing an observation protocol to engineering related learning.

6.4 The Importance of Observation Protocols in Early Engineering

Child development and its applications in early childhood education and care were created beginning in the 1920s, in the Western world, through the collection of long and systematic observations. Early psychological studies of very young children also made use of the observational method as a window into a child's "mind". Observational methods are well suited for early childhood education research and practice and have traditionally been used to produce reliable findings leading to good practice (Jamblon, Dombro, & Dichtelmiller, 2007). These empirical methods are appropriate because they permit direct access to recording children's behaviors at a time in development where language offers limited indications of thinking and intent.

In regard to early engineering, establishing an area of research study requires good definitions of the phenomenon under study that are reliably documented and in methodologically accepted ways. When the, now very famous, programs for young children in the Regio Emilia district attempted to first distill the originality, depth, and artistic ingenuity of the children participating in their programs, they did so through observation (Rinaldi, 2006a, b). In the same tradition, developmental engineering as it maps child behaviors that are predecessors to engineering thinking must make use of the observational approach at first. The PREEOP is a protocol that has resulted from extensive observation through the early engineering lens. It is intended to be used in the classroom by a teacher, while lively child-centered and developmentally appropriate curriculum guides children's behavior.

In addition to using observational methods to document children's development within a specific context, in our case engineering, these direct ways of "looking" at children have often lead to the development of modern ideas on what constitutes good learning for very young children. The Pre-Kindergarten Engineering Observation Protocol (PREEOP) fits well into that long tradition, and compliments contemporary ideas on the content of early childhood education. Specifically, the trend to establish observational methods is at the heart of not only understanding what children do but also builds curriculum around these actions and their interpretations otherwise known as Emergent Curriculum (Stacey, 2009). Tools like the PREOP create an opportunity for practitioners to guide their observation outside commonly understood parameters of child action like pouring water or stacking blocks and categorizing these actions under a new formalization in the engineering sciences. Emergent curriculum scholars like Rinaldi (2006a, b) and Wien (1995) argue convincingly that our early childhood education curricula are only as good as our observation methods, protocols, and skills permit.

6.4.1 The Protocol

The PREEOP protocol has been designed based on findings from our initial studies as well as STEM-related learning identified through the P2E classroom based curriculum (Bagiati, 2011; Bagiati and Evangelou, 2015; Bagiati, Yoon, Evangelou, & Ngambeki, 2010; Bagiati, Yoon, Evangelou, Magana, et al., 2015; Bagiati & Evangelou, 2016; Bairaktarova et al., 2011; Evangelou et al., 2010). Use of the PREEOP in a classroom is expected to enable a teacher to identify learning in the four categories defined by Katz (1999), namely knowledge, skills, dispositions and feelings, through a STEM lens. To provide a better understanding, examples of identified learning instances have been provided. At this point we should also remind the reader that through all our studies children were able to express STEM-related learning both through declaration as well as through mere demonstration, therefore both conditions should be observed and taken into account when the PREEOP is used (Tables 6.1, 6.2, 6.3 and 6.4).

6.5 Conclusions

In this paper we contribute both conceptually and methodologically in presenting an observational tool that can establish the validity of the early engineering concept within early education. Observation is often the starting point in seeking to describe newly conceived concepts as well as a methodological tool frequently employed when emphasis is primarily on action. Observing and recording child behavior in contexts of engineering praxis has led us to the development of behavioral categories that are measurable and verifiable and that provide a platform for further development of research tools. Further discussion on the assessment of early engineering

Table 6.1 PREEOP—STEM knowledge

STEM—knowledge subcategory	Demonstrated learning instance	Examples
Math	The child demonstrates knowledge relevant to numeracy, geometry, scale, ratio, and space concepts	"Look I have three cars and a dog"
Science	The child demonstrates knowledge of content relevant to weather phenomena, biology, and physics	"It's cloudy, so it's going to rain"
Technology	The child demonstrates knowledge relevant to the existence or use of technology	"This is a pick-up truck. I can lift boxes"
Artifacts	The child demonstrates knowledge relevant to manmade artifacts	"I helped my parents build a table"
Functions	The child demonstrates knowledge relevant to the fact that manmade artifacts and constructions serve a teleological purpose	"I will use the phone to call grandma"
Buildings	The child demonstrates knowledge relevant to the existence of building constructions, or particular elements of these constructions	"We put windows in the houses so we can see out"
Materials	The child demonstrates knowledge relevant to the existence and the attributes of materials	"I want to use bricks to build my house but they are heavy!"
Construction	The child demonstrates knowledge relevant to the fact that materials go through a construction process to become functional artifacts	"We cannot use it now. We have to let it dry first"
Design mental models	The child demonstrates existence of a mental model regarding a specific design to be implemented. The design mental model may refer to function or structure	Child while drawing a parking garage "My cars will go in from here, then go up, more up, and then go out from here"
Engineering design process	The child demonstrates knowledge relevant to the engineering design process or to its phases	"We can go to a fire station and look for more ideas!"
Engineer profession	The child demonstrates knowledge about the engineering profession	"Engineers build boats and houses and phones"
Problem solving	The child demonstrates knowledge about the existence of a conceptual or structural problem, or demonstrates knowledge regarding an attempted or implemented solution	"My car is not turning. I need to make a new steering wheel"

thinking and development appears in the next chapter, where a model that guides both assessment practices and practical teaching strategies is presented.

The studies presented in this chapter suggest that research in early engineering has the capacity to contribute to our understanding of significant questions within the purview of STEM in general and engineering education in particular. Specifically, early engineering can inform efforts to push school curricula towards higher integration starting with the early years. Engineering as a practical field provides

Table 6.2 PREEOP—STEM skills

STEM—skill subcategory	Demonstrated learning instance	Examples
Communicating	Child demonstrates a type of communication related to the early engineering project. Communication may be verbal or non-verbal and it may be between the child and an adult or between two or more children	Teacher: "Where could Sam's dolls live?" Child: "The dolls can live in a playground!"
Reasoning	Child demonstrates ability to support his/her arguments and to draw inferences	Teacher: "How do you know what's in the box?" Child: "I can smell it!"
Questioning	Child demonstrates ability to generate questions related to the early engineering project	"Why don't we make it bigger?"
Setting goals	Child demonstrate the ability to discuss or work towards a goal/purpose and communicates the goal while talking, constructing or using materials	"I want to build a house for my dog to sleep in"
Drawing	Child demonstrates ability to draw an artifact, building, or part of the environment	"Look I draw a school!"
Designing/constructing	Child demonstrates ability to design/construct an artifact or building that serves a particular purpose	"I am building a street for my car"
Problem Solving	Child demonstrates ability to solve a problem. The solution to the problem may be implemented by the child or be verbally proposed by the child to another child or an adult	"Now it's too short but we can put it on a block to make it tall"
Synthesizing	Child demonstrates ability to combine his/her construction with someone else's construction and create a common final one. Furthermore, Child demonstrates ability to use different types of resources available in the room that would initially belong to different toy categories, to create a final construction project	John could not make his tree stand, but then he went to the kitchen area picked up some popsicle sticks and used them to support the structure
Collaborating	Child demonstrates ability or the willingness to collaborate with an adult or with one or more children to complete a task	"Can you hold this for me to glue? I can not do it alone"
Comparing	Child demonstrates ability to perform a comparison between two situations before producing an inference	"His house is taller but I can fit more dolls in mine!"
Testing/evaluating	Child stops constructing to test the construction and to evaluate whether the object functions as needed or planed	"My tree is not standing. I need to make it stand"
Elaborating/improving	Child demonstrates ability to further elaborate on a prior design and identify one feature that has improved	"I want to make my house bigger today to fit my car"
Reproducing	Child demonstrates the ability to redo, redesign or reconstruct the whole or part of something he/she or another child has created recently or in the past	"No we did not build it like that! We used the big block!
Giving or following instructions	Child demonstrates ability to give or follow step by step instructions to complete a task	"Put it out of the window so Sam can see it!"
Explaining how things are/work	Child explains during or at the end of his/her activity what he/she is making or what he/she has done	"You have to lift this and press the button"
Use of Technology	Child demonstrates ability to use technology on his/her own	"My mom lets me click the mouse and print my photo"
Use of technical language	Child demonstrates the ability to accurately use technical language	"We can fix it with a screwdriver"

Table 6.3 PREEOP—STEM dispositions

STEM—disposition subcategories	Demonstrated learning instance	Examples
Construction/building	Child demonstrates disposition to construct things or to build in the block building area	Mary has build 4 different buildings within a week
Problem solving	Child demonstrates disposition to participate in problem solving discussions or activities	George loves helping other kids fix toys/constructions that don't work
Particular materials	Child demonstrates disposition to use or propose particular materials to be used in his/her designs	Nick's drawings always present buildings made of bricks
Particular themes	Child demonstrates disposition to incorporate particular themes in his/her work	Susan always likes to draw/build garages
Particular collaborations	Child demonstrates disposition to perform his/her work or select his/her free play activities based on collaborating with one or more particular children	Kate, John and Mary always like to build together
Creativity & Innovation	Child demonstrates disposition to presents a design, in the form of a drawing, a maquette, or a block construction that has never been introduced in the classroom before	Dave always builds cars, but it's a different type of car every time
Acknowledgment	Child demonstrates disposition towards having his/her work acknowledged/documented	Monique always wants the teacher to put her drawing on the wall when she is done

the appropriate framework within which early education curricula can explore the enhancement of special skills and knowledge that are both developmentally appropriate and socially desirable for the 21st-century learners. Our studies indicate that this is possible as well as desirable but a lot of foundational work remains to be done. Defining and refining research categories, utilizing the design cycle and exploring the boundaries could open early education curricula and help us identify the nexus of useful concepts in early engineering education.

During this attempt, special attention has to be directed to issues of authenticity both in terms of early education curricula as well as engineering content. Any attempt to bring engineering into early education must be accompanied by sincere and systematic efforts to educate teachers so that they may take on the responsibility of introducing engineering in their classrooms effectively and with confidence. In that direction, a lot remains to be done as pedagogical systems and teacher education tend to be slow in adopting innovation.

Table 6.4 PREEOP—STEM feelings

STEM—feelings subcategories	Demonstrated learning instance	Examples
Enthusiasm	The child appears to be enthusiastic and having a good time during the early engineering class discussions or activities	"Today I'm going to build a fire station!!"
Boredom/distraction	The child expresses a desire to leave the current early engineering discussion/activity to proceed with free play; expresses the desire to not participate in an early engineering discussion/activity proposed to him/her; or is reported by the observer to be distracted	"I don't want to draw for Sam's toys anymore, I want to read my book"
Pride by self-achievement	The child appears to be happy or proud of his/her own achievements or his/her own ideas	"I got a great idea! Why don't we make it red!"
Frustration by failure	The child appears to be frustrated by his/her own inability to successfully complete the task he/she was engaged into	"I just try and try and it does not stay!
Pleasure by working with parents	The child appears to be pleased by the parents' involvement in the early engineering project at home or in the classroom	"My mom helped me build it!"
Pleasure by collaborations	The child appears to be pleased by the collaborating with particular adults/children during the early engineering project	"Hey look what we built!! It's a castle!
Frustration by collaborations	The child appears to be frustrated by someone else's design idea, use of resources, or design implementation	"Stop kicking my street!!
Engagement	The child appears to be interested and engaged to the P2E discussions or activities	"Can I draw a little more? I have a new idea!"
Liking of resources	The child appears to be happy because he/she likes the resources used within the	"I like the colors on these new blocks a lot!"
Dislike of resources	The child expresses a desire to leave the current early engineering activity or expresses the desire to not participate in a early engineering activity proposed to him/her, because he/she does not like the resources used in the activity	"I don't want to do it. My duck tape never sticks!

It is important to keep in mind early engineering differs significantly from other kinds of engineering education directed at higher levels of schooling. Early engineering seeks to understand something about the fundamental nature of human activity that is universally recognized and understood in its role of stewardship of the human made world. Furthermore, it is important to understand that we should not incorporate engineering in early childhood only as a way to prepare skillful learners but we must also seek ways to benefit from its effects on cultivating creativity and general problem-solving ability.

Early engineering has the capacity to usher a new era of methodological innovation as it is compelled to bridge such conceptually distal areas. Imagining, creating, verifying and establishing the commonalities between human universals at the onset of early behavior and the human creativity par excellence, as is engineering, is a daunting task. It is also a joyful task as it stretches our known boundaries and seeks new borders.

Appendix 1: Terminology and Concepts Addressed Through P2E Lesson Plans (Bagiati, 2011, pp. 60–62)

Lesson plan title	New engineering-related concepts and terminology	Revisited engineering concepts and terminology	Rationale to be addressed
"I have a great idea!"	Engineers, desire, discussion, usage		Engineering usually is initiated by **someone's desire** to create and construct something new or to alter something that already exists in order to achieve a new goal. Engineers **discuss** the new idea about how things can be **used**.
"Making a decision!"	Decision-making, brainstorming, constraints, criteria	Engineers, discussion, usage	Engineers usually **brainstorm** and then revisit their initial ideas to see if any new idea has appeared before coming to a **final decision**. After brainstorming and coming up with a set of new ideas or possible solutions, engineers have to also consider the **constraints**.

Lesson plan title	New engineering-related concepts and terminology	Revisited engineering concepts and terminology	Rationale to be addressed
"Let's look at building elements!"	Building, observing, elements	Engineers, discussion, usage, decision making, criteria	While **constructing a building** there are plenty of building **elements** that an engineer can think about designing and implementing.
"Let's look what it is made of!"	Materials	Building, observing, elements, decision making, criteria	While constructing a **building**, depending on the climate, cost, usability, and numerous other factors, an **engineer** can **select** to use different **materials** on the building.
"Let's make a model"	Design, model	Building, elements, materials	After brainstorming and doing some first drafts on paper, engineers start to try to come up with **3D models** of the **construction**, which can be made out of many **different materials**: paper models, cardboard models, and models with mixed materials.
"Let's improve our models"	Improve	Discussion, design, model, material, usage	Today you will revisit the **3D models** you created, **discuss** them, and attempt to **improve** them.
"Let's show our ideas"	Design representations, drawing, sketch, floor plan	Engineers, design, models	Engineers are using **different representations** in order to communicate their ideas with clients/other engineers/constructors, etc. Today you will introduce children to these types of representations.

Lesson plan title	New engineering-related concepts and terminology	Revisited engineering concepts and terminology	Rationale to be addressed
"Let's show our ideas differently"	Maquettes, consult	Design representations, drawing, sketch, floor plan, model, improve	Today children will revisit and discuss these **representations** and work on their **maquettes.**
"Let's sketch it"	Sketch, present	Buildings, observing, drawing, discussion	There is a lot of discussion regarding how much inspiration and creativity freehand sketching is stimulating. For this reason, we want to take the children on a field trip in order to **observe** and **free sketch** their environment.
"Let's observe our building"	Electricity, water, buttons, switches, pipes, function	Building, sketch, present, observing, discussion, usage	**Discuss** the **sketches**. Let the children **present** them. Then start **observing** and identifying engineering features in the school building, and start making connections between **action and effect**(e.g. pressing buttons).
"Let's investigate some more"	Handles, investigate	Building, discussion, electricity, water, buttons, switches, pipes, function, observing	Keep **observing** the **building**. Add one more element for **investigation** and **discussion** (i.e., **handles**).
"Let's see what is around us"	Tables, graphs	Buttons, switches, pipes, handles	After **gathering information**, engineers use various **visual representation tools** in order to make the results of the data gathered more obvious. Today the children will do **tables** and **graphs**.

Lesson plan title	New engineering-related concepts and terminology	Revisited engineering concepts and terminology	Rationale to be addressed
"Let's get prepared for our constructions"	Information gathering	Criteria list, materials	After having come up with some initial **ideas**, engineers are creating **criteria/requirements** lists in order to start planning the actual **construction**. This is also a good time to **consult** other engineers in order to **discuss problems** that may appear in the design, or just to get more ideas.
"Let's sketch some more"		Sketch, observe, present	The children will go on another field trip in order to **observe** and **free sketch** their environment.
"Let's observe our surroundings"	Get inspired	Observe, discuss	In today's class, we want the children to **observe engineering features** in their surroundings that may be used as **inspiration** to their **project**.
"Let's see where the light is"	Electricity, circuit, lamp	Sketches, drawings, switch, maquettes, model, improve, test	When coming closer to an end product, engineers start to bring more **details** in their **maquettes** and **models**. Revisit previous representation types (e.g., **drawings, sketches, maquettes**), and try to add more features to them. Today children will create simple **electric circuits** and will be prompted to use them within their previous work.

Lesson plan title	New engineering-related concepts and terminology	Revisited engineering concepts and terminology	Rationale to be addressed
"Let's search a little more"		Information gathering, communication, consulting, presenting	When coming closer to an end product, **engineers** do more detailed **information gathering** about particular **elements** of the **design**. Information gathering can include looking at books, online searching, and discussing the issue with other **experts** in the field. When new ideas are on the table, engineers have to explain the new details to other members of their team. Some days ago, parents were asked to conduct research regarding the project with their children at home or in the surrounding area, and send the information with their children to class. Today the children will **present** their findings.
"Let's pay a visit"		Consulting, information gathering, observe, communication, exchange of ideas	Today the children will go on a field trip to visit a facility related to their project and **gather more information** about it by **talking** to the people who work there.

Lesson plan title	New engineering-related concepts and terminology	Revisited engineering concepts and terminology	Rationale to be addressed
"Let's ask an engineer"		Model, consulting, improving, comparing, testing, construction, electric circuits, buildings	While developing a final **model**, engineers **try, test, and compare** different solutions for various **elements** in their **construction**. Along with their personal testing, **consulting** also takes place. It is not necessary that only one solution is the optimum every time. Today parents or friends that are engineers will visit the class to show the children different ways to assemble and test **electric circuits** and to discuss the children's **building** ideas.
"Let's see what we can build with"		Model, consulting, information gathering, communication, exchange of ideas, materials, presentation	Elements have to be selected to be used for the **final model** construction to resemble reality and represent the **design** features as beneficial as possible. Today the children will **present** in class **building materials** they brought from home, and they will discuss how they could use them to build their final **constructions**. Parents and friends that are engineers have also been invited to be in class and participate in the **discussion** and the **building process**.

Lesson plan title	New engineering-related concepts and terminology	Revisited engineering concepts and terminology	Rationale to be addressed
"Let's see how we can improve it"	Decorate	Test, improve, model	Upon completion of construction of the **model**, engineers revisit it to **test and improve** it even further, and then start to add **decorative elements** to provide more context.
"Let's build our village"	Synthesize, finalize,	Model, decorate	Upon completion of construction of the **final model**, engineers revisit it to **test and improve** the final deliverable and add **decorative** elements to provide more context.
"Let's show it to our friends"	Explain, invite	Model, engineers, present	Upon completion of construction of the **final model**, engineers **present** them to colleagues or customers. Today the children will present their work to children from other classrooms, and they will create invitations to invite their parents to come and see their work.
"Let's show it to our parents"		Model, maquette engineers, present	Today the children will **present** their work to their parents.

References

ABET. (2016). *Criteria for accrediting engineering programs, 2016–2017*. Retrieved from http://www.abet.org/accreditation/accreditation-criteria/criteria-for-accrediting-engineering-programs-2016-2017/.

Bagiati, A (2011). Early engineering: A developmentally appropriate curriculum for young children. *Dissertation, School of Engineering Education*. Purdue University.

Bagiati, A., & Evangelou, D. (2011, October 4–7). Starting young: Outcomes of a developmentally appropriate PreK engineering curriculum. In *Research in Engineering Education Symposium, Madrid, Spain*.

Bagiati, A., Evangelou, D. (2015). Engineering curriculum in the preschool classroom: The teacher's experience. *European Early Childhood Education Research Journal, 23*(1).

Bagiati, A., & Evangelou, D. (2016). Practicing engineering while building with blocks: Identifying engineering thinking. *European Early Childhood Education Research Journal,* 24(1).

Bagiati, A., Yoon Yoon, S., Evangelou, D., Kaloustian, G., Cekic, O., Zhu, J., & Magana, A. (2011, June 26–29). Engineering in early education: A Multicultural comparison of web resources. In *Proceedings of the American Society for Engineering Education Annual Conference, Vancouver, Canada.*

Bagiati, A., Yoon, S.Y., Evangelou, D., & Ngambeki, I. (2010). Engineering curricula in early education: Describing the landscape of open resources. *Early Childhood Research & Practice, V12*(2).

Bagiati, A., Yoon, S.Y., Evangelou, D., Magana, A., Kaloustian, G., & Zhu, J. (2015). The landscape of PreK-12 engineering online resources for teachers: Global trends. *International Journal of STEM Education, V2*(1).

Bairaktarova, D., Evangelou, D., Bagiati, A., & Brophy, S. (2011). Engineering in young children's exploratory play with tangible materials. *Children, Youth and Environments, 21*(2).

Bonawitz, E., Shafto, P., Gweon, H., Chang, I., Katz, S., & Schulz, L. (2009). The double-edged sword of pedagogy: Modeling the effect of pedagogical contexts on preschoolers' exploratory play. In *Proceedings of the 31st Annual Meeting of the Cognitive Science Society, July 29th-August 1st.* Netherlands: VU University Amsterdam.

Brophy, S., & Evangelou, D. (2007). Precursors to Engineering Thinking (PET) Project: Intentional Designs with Experimental Artifacts (IDEA). In *Proceedings of the American Society for Engineering Education Annual Conference and Exposition*. Honolulu, HI: American Society for Engineering Education.

Brophy, S., Klein, S., Portsmore, M., & Rogers, C. (2008). Advancing engineering education in P-12 classrooms. *Journal of Engineering Education, 97*(3), 369.

Cook, C., Goodman, N. D., & Schulz, L. E. (2011). Where science starts: Spontaneous experiments in preschoolers' exploratory play. *Cognition, 120*(3), 341–349.

Culver, D. E. (2012). *A qualitative assessment of preservice elementary teachers' formative perceptions regarding engineering and K-12 engineering education* (Master's thesis). Retrieved from http://lib.dr.iastate.edu/etd. (Paper 12888).

DeJarnette, N. (2012). America's children: Providing early exposure to STEM (science, technology, engineering and math) initiatives. *Education, 133*(1), 77–84.

DiFrancesca, D., Lee, C., & McIntyre, E. (2014). Where Is the "E" in STEM for young children? Engineering Design education in an elementary teacher preparation program. *Issues in Teacher Education, 23*(1), 49.

Dodge, D. T. & Colker, L. (1996). The creative curriculum for early childhood (3rd ed.).

Duncan, D., Diefes-Dux, H., & Gentry, M. (2011). Professional development through engineering academies: An examination of elementary teachers' recognition and understanding of engineering. *Journal of Engineering Education, 100*(3), 520–539.

Evangelou, D., Dobbs-Oates, J., Bagiati, A., Liang, S., & Choi, J. Y. (2010). Talking about artifacts: Preschool children's explorations with sketches, stories, and tangible objects. *Early Childhood Research & Practice*, 12(2).

Hedtke, R., Kahlert, J., & Schwier, V. (2001). Service industry for teachers? Using the Internet to plan lessons. *European Journal of Education, 36*(2), 189–193.

Hsu, M. C., Purzer, S., & Cardella, M. E. (2011). Elementary teachers' views about teaching design, engineering, and technology. *Journal of Pre-College Engineering Education Research (J-PEER), 1*(2), 31–39.

Jahan, K., & DeJarnette, N. (2014). A mobile program for enhancing engineering education. In *2014 IEEE Frontiers in Education Conference (FIE) Proceedings* (pp. 1–5). IEEE.

Jamblon, J. R, Dombro, A. L., & Dichtelmiller, M. L. (2007). *The power of observation: Birth through eight* (2nd ed.). Teaching Strategies. ISBN 9781933021.

Jennings, K. D., Harmon, R. J., Morgan, G. A., Gaiter, J. L., & Yarrow, L. J. (1979). Exploratory play as an index of mastery motivation: Relationships to persistence, cognitive functioning, and environmental measures. *Developmental Psychology, 15*(4), 386.

Katehi, L., Pearson, G., & Feder, M. (2009). *Engineering in K-12 education: Understanding the status and improving the prospects* (594p). Performer: National Academy of Engineering, Washington, DC.; National Academy of Sciences-National Research Council, Washington, D.C.

Katz, L. G. (1999). *Another look at what young children should be learning*. ERIC Digest. Available: http://files.eric.ed.gov/fulltext/ED430735.pdf.

Katz, L. G., & Chard, S. (2000). Engaging children's minds: The project approach (2nd ed.).

Korošec, H. (2012). Playing with puppets in class—Teaching and learning with pleasure. *The Power of the Puppet*, 29.

Majaron, E. (2012). Arts as the pathway to the child. *The Power of the Puppet,* 11–18.

National Association for the Education of Young Children. (2009). *Developmentally Appropriate practice in early childhood programs serving children from birth through age 8*. Position Statement: National Association for the Education of Young Children.

Next Generation Science Standards. (2013). *Next generation science standards: For states by states, next generation science standards*. Retrieved from https://www.nap.edu/login.php%3frecord_id%3d18290%26page%3dhttps%3A%2F%2Fwww.nap.edu%2Fdownload%2F18290.

Newburger, A., & Vaughan, E. (2006). *Teaching numeracy, langage, and literacy with blocks*. St Paul, MN: Redleaf Press.

Piaget, J. (1929). *The child's conception of the world*. New York: Harcort, Brace.

Recker, M. (2006). Perspectives on teachers as digital library users: Consumers, contributors, and designers. *D-Lib Magazine, 12*(9), 2.

Rinaldi, C. (2006a). *In dialogue with Regio Emilia: Listening, researching, and learning*. New York: Routledge.

Rinaldi, C. (2006b). *In dialogue with Reggio Emilia: Listening, researching and learning*. Psychology Press.

Stacey, S. (2009). *Emergent curriculum in early childhood settings: From theory to practice*. Saint Paul, MN: Redleaf Press.

Tepylo, D. H., Moss, J., & Stephenson, C. (2015). *A developmental look at a rigorous block play program. Young Children, 18–25*. Retrieved from http://www.naeyc.org/yc/files/yc/file/201503/YC0315_Tepylo.pdf.

Wien, C. A. (1995). *Developmentally appropriate practice in "Real Life": Stories of teacher practical knowledge*. New York: Teachers College Press.

Chapter 7
Assessing Early Engineering Thinking and Design Competencies in the Classroom

Şenay Purzer and Kerrie Anna Douglas

Abstract Young children are capable of understanding ideas that educators had once thought to be too complex for these ages. Children start to engage in creative design and develop engineering thinking at early ages as they play, create, solve puzzles, and ask questions. Just as it is important to highlight these activities as early engineering practices, it is important to use assessment practices necessary to support further development of engineering thinking. In this chapter, we lay the foundation for assessment of young children's engineering thinking through discussion of current research on early engineering thinking and effective approaches to assessment as we outline engineering design competencies for young learners. We also present the Mosaic framework, a model that guides assessment practices in engineering and provides practical strategies that are necessary to maintain complexity while teaching and assessing engineering design to young children. We urge the community toward a multi-faceted view of assessment that targets student learning evidence and growth supported by curriculum design, and teacher professional development, along with assessment tools and strategies.

7.1 Introduction

Researchers and educators are at an exciting and interesting crossroads with regard to the assessment of engineering thinking in early childhood education. In the past several decades, many developments have taken place in the USA with implications for assessment, early childhood education, and engineering education. In the field

Ş. Purzer (✉) · K. A. Douglas
School of Engineering Education, Purdue University, West Lafayette, IN, USA
e-mail: purzer@purdue.edu

K. A. Douglas
e-mail: douglask@purdue.edu

L. English and T. Moore (eds.), *Early Engineering Learning*, Early Mathematics Learning and Development, https://doi.org/10.1007/978-981-10-8621-2_7

of assessment, reform efforts have promoted a greater emphasis on identifying what students can do, rather than simply what they can memorize and recite (Pellegrino, 2012). In essence, educational reform has necessitated the assessment of higher-level thinking skills, such as critical thinking and creativity. In the field of early childhood education, researchers call for the integration of research on children's cognition and learning processes into instruction and a better preparation of educators and caregivers in research-informed teaching and assessment practices (Institute of Medicine (IOM) & National Research Council (NRC), 2012). Furthermore, the NRC report, *Eager to Learn: Educating our Preschoolers*, asserts that assessment has an "important role to play in revealing a child's prior knowledge, development of concepts, and ways of interacting with and understanding the world so that teachers can choose a pedagogical approach and curricular materials that will support the child's further learning and development" (NRC, 2000, p. 259). Finally, engineering and integrated STEM competencies have gained a greater importance at all levels of education since 2009 with the publication of the *Engineering in K-12 Education: Understanding the Status and Improving the Prospects* by NAE and NRC (2009). The policy changes in engineering education enabled a greater emphasis on higher-level thinking skills such as argumentation and decision making while engaging children in open-ended problem solving.

These developments from three different fronts put preschoolers, kindergarteners, and primary school children and educators at the center of a crossroads urging for transformation at the early stages of the public education system. Children go through a significant and often irregular rate of growth during the first five years of their life (NRC, 2000). In these years, children develop many skills ranging from sharing to counting. There is, hence, an opportunity to scaffold young children in early engineering thinking and design as foundations for life-long critical thinking and creativity. As several chapters in this book have illustrated, this learning must be carefully captured, further scaffolded, and housed within supportive learning environments. Such rapid growth of the young brain makes it even more important that educators and caregivers are trained to engage in effective assessment and scaffolding with questioning, observing, and many other means.

The consistent message across the fields of assessment, early childhood education, and engineering via integrated STEM education is that higher-levels of thinking and skill development must be supported with high-quality assessment processes that inform teaching and learning. As this book indicates, children start to engage in creative design and develop engineering thinking at early ages as they play, create, solve puzzles, and ask questions. However, these activities often are not labeled as engineering. Engaging in engineering design promotes critical capabilities in early childhood education. In this chapter, we aim to untangle these diverse dimensions of learning and assessment as part of an ecosystem that involves curriculum design, teacher professional development, and assessment tools and strategies. We specifically focus on the use of assessment to promote engineering design competencies among young children. We emphasize the importance of focusing on evidence of student learning and growth in an area that we are most passionate about.

7.2 Early Engineering Thinking

Engineering and design education has been a topic of interest globally in the last twenty years. More recently in the USA, there has been increased attention to research and policy that bring engineering into pre-college settings with greater emphasis than it has been before (NAE & NRC, 2009; Purzer, Strobel, & Cardella, 2014). These developments provide the motivation for thinking more clearly about assessment and an opportunity for understanding engineering thinking in young children. As a field focused on problem solving, engineering builds on and integrates aspects of the other STEM disciplines: science, technology, and mathematics.

In engineering, design is the overarching process of inquiry into solving problems (Atman & Adams, 2007; Daly, Adams, & Bodner, 2012; Sheppard, Macatangay, Colby, & Sullivan, 2008). Design problems, however, are not straightforward, well-defined problems. Engineering involves solving problems with incomplete information (Gainsburg, 2006) and implicit goals that need to be uncovered by the designer (Jonassen, Strobel, & Lee, 2006). Moreover, there are no single, correct solutions, rather a variety of solutions that sufficiently meet the needs of diverse stakeholders based on the designer's approach to trade-offs. However, although design is a complex multi-faceted process in engineering practice, it is also accessible for young children (Hsu, Cardella, & Purzer, 2012), as the chapters in this section have illustrated. For example, in Fig. 7.1, we present the sketch of a barn that a four-year old has drawn to meet the needs of her toy cow. In this example, the child is recognizing the basic needs of the animal (water and food) but also includes other elements such as milk and eggs. Design challenges, when presented in familiar and engaging contexts, can help elicit children's prior knowledge through sketches or three-dimensional models and explanations children articulate through teacher questioning (e.g., What do cows need to live? What are the eggs for? Is the milk for the cows?).

7.3 Research on Young Children's Engineering Thinking

Children are naturally good at imagining, building, testing, and improving. These are also common practices in engineering design. Hence, children naturally engage in design when playing (Stagnitti, 2014). With these built-in competencies, children can benefit from engaging in problems that they perceive to be authentic but unfamiliar. When designed carefully, engineering design projects can allow children autonomy while providing multiple means of support as they do so for adult learners (Purzer & Fernandez, 2015). For example, an authentic problem can be the challenge of designing a solution that allows a two-year old to easily dispense a small amount of toothpaste while brushing. Another challenge that early childhood students would find authentic is designing for a character in a book (Novel Engineering, 2016), from a cartoon show, or a toy they regularly play with. While the ideas around design

Fig. 7.1 Design sketch of a barn for a toy cow by a four-year old (note: green writing is researcher text added to make clear the parts drawn within the image)

related to imagining, building, testing, and improving might be new to educators, these ideas are not new to children.

Educators also agree on the importance of questioning and the necessity in supporting children's abilities to ask questions (Hunter & Sonter, 2011). Although we might assume that children are natural at asking questions, children do not naturally associate asking and planning as critical aspects of design problem solving. Hsu, Cardella, and Purzer (2012) conducted a study using an interview protocol to assess students' understanding of the engineering design process using an illustration of a character designing a container for an egg-drop contest. They asked participant students to describe what they thought was good about the process illustrated and what they would do differently. This cross-sectional study showed that, while elementary students' understanding of the engineering design process progressed over the years even without any formal engineering instruction, asking and planning were the most difficult for students to recognize (see Fig. 7.2). This finding is interestingly in congruence with other studies involving college students (Atman, Cardella, Turns, & Adams, 2005). Their analysis revealed that students discuss "testing" as the only method of finding out if a design solution works.

7.4 Engineering Requires Both Creative and Critical Thinking

Engineering provides meaningful and engaging environments for children to become scientifically and technologically literate. At the heart of engineering is creative

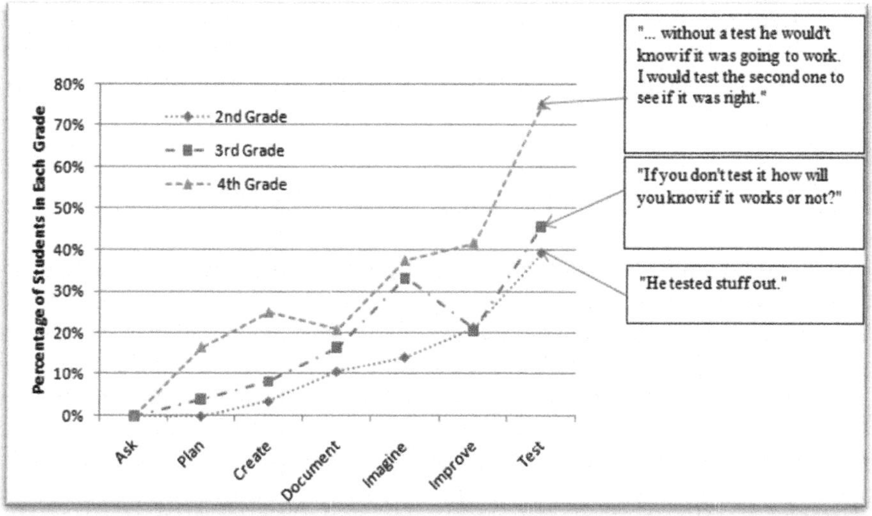

Fig. 7.2 A cross-sectional comparison of elementary students' understanding of design (from Hsu, Cardella, & Purzer, 2010)

thinking (divergent thinking) and critical thinking (convergent thinking). As part of early elementary students' STEM literacy, it is important to develop their divergent and convergent thinking, and engineering is one vehicle to accomplish this.

Creative thinking includes generating ideas fluently, elaborating on ideas, and associating ideas in novel ways. In situations that require association, a child can, for example, associate what a house is for people and what a burrow is for hamsters or ants (PictureSTEM, 2016a). A child, when given a compass, can describe his or her observations and list ways this device can be used in different ways. Children's limited experiences with instruments can be an opportunity to tap into their creative, divergent thinking. In these novel experiences, young children also can demonstrate their ability to make new associations and inferences from these associations.

Critical thinking includes analyzing, comparing, prioritizing, questioning, and making judgments which can be addressed with early elementary learners. For example, children aged five to six can predict and explore fundamental engineering science concepts such as material strength and failure by designing baskets using different materials (PictureSTEM, 2016b) and explore friction on an inclined plane (ETA hand2mind, 2013). We anticipate that young children, if given opportunity and support, can also tackle even more complex problems and develop potential solutions. For example, when considering the problem of mosquito infestations, children can explore different animals that eat mosquitos (e.g., bats and sparrows) and compare habitats these animals live in. Moreover, when introduced in ways they find meaningful and engaging, children can also learn specific scientific and technical terms such as nocturnal (to describe specific attributes of animals such as bats) and biopesticides (to describe an approach taken to solve the problem of mosquito infestation).

Engineering contexts can foster young learners' creative thinking by allowing them the flexibility and opportunity to think in new ways and critical thinking by providing scaffolding and support to help them make decisions about their ideas.

7.5 Assessment of Early Engineering Thinking

The two overarching aspects of design that tap into children's creative thinking and critical thinking abilities have implications for assessment of young learners. For example, teaching engineering in early childhood education necessitates assessment that can capture children's creative practices along with their reasoning and decision-making processes. A key challenge in assessment of design and engineering thinking, particularly among children, is maintaining complexity and wholeness of the design practice while making it practically accessible to children as well as their teachers.

Earlier, we have discussed that design problems are ill-defined (Jonassen et al., 2006). There are no single, correct solutions, rather a variety of solutions that sufficiently meet the needs of the users or a set of stakeholders. Premature scaffolding of these aspects of design can limit opportunities for learning. Simplification can lead to design challenges that have no context. Often in these types of design challenges, the information is given to the children without engaging children's curiosity to ask for information. In examples where the complexity of design practices is maintained, children are expected to retrieve their prior knowledge and gather new information to help answer emerging questions. In Table 7.1, we present example activities within a design challenge where students are designing a barn for a cow. Within the stages of engineering design, the design practices range from oversimplified to appropriately complex. The oversimplified versions of design practices lack a direct user, have clear design criteria and constraints, and urge the child toward a single solution. The design practices that are appropriately complex can engage children in both creative and critical thinking as they uncover the needs of users and learn to justify their solutions rather than merely describe them.

7.6 A Framework for Assessing Engineering Thinking and Design Competencies

Assessment allows for inquiry into children's thinking. It is our way of collecting, analyzing, and judging records of evidence that provide insights on children's learning as well as emotional health and social development. Moreover, assessment is an inseparable part of the teaching process that can help reinforce children's development of higher levels of thinking and skills. Here, we envision children developing higher-level learning in environments focused on engineering problem solving by building on the premises of the other STEM disciplines: science, technology, and

Table 7.1 Variations of design practices for children

	Examples of design challenges Oversimplified ←………………… → Appropriately complex		
Design challenge	Build a barn	Build a barn for your toy cow	A farmer has donated a cow to our city zoo. The zoo needs to find a place for this new cow
Information gathering	Given: Animals need water, food, and shelter	Explored: What does your animal need?	Explored: What do we know about this animal? What questions do you have?
Uncovering goals (criteria and constraints)	Given: Build a barn that is big enough for your cow and has space for food and water	Given: Build a barn that is big enough for your cow and has space for food and water	Scaffolded: What are the needs of this animal?
Solution generation	Generate a solution	Generate two or more solutions	Generate at least four solutions that we can compare
Justifying solutions (decision making)	Explain your solution	Scaffolded: Is your barn big enough for your cow and has space for food and water?	Describe how your solution meets the needs of the animal. Justify how your solutions meet design criteria and constraints. Share any trade-offs you made.

mathematics. What we envision can be considered to be too complex to put into practice. Hence, in order to frame components of this complex learning environment in a way to support effective teaching, learning, and assessment, we developed a framework, called Mosaic. The Mosaic Framework is inspired by the metaphor for assembling small pieces to create a meaningful expression and has three main components: a base, a mortar, and tiles (see Fig. 7.3).

Mosaic's base is made up of essential goals and objectives aligned with high-quality curriculum, instruction, and embedded assessment tools and strategies. Mosaic's mortar keeps base and tiles together through support mechanisms such as teacher professional development. The tiles consist of assessment tasks that target aspects of learning embedded within a curriculum and transfer abilities of the children. Each component of the Mosaic also represents a set of key principles which are highlighted in the next section. With this framework, we argue that teachers who are knowledgeable about the performance of their students that target high levels of learning in complex engineering design environments and able to use evidence from student performances to support student learning, will foster students' critical and creative thinking skills.

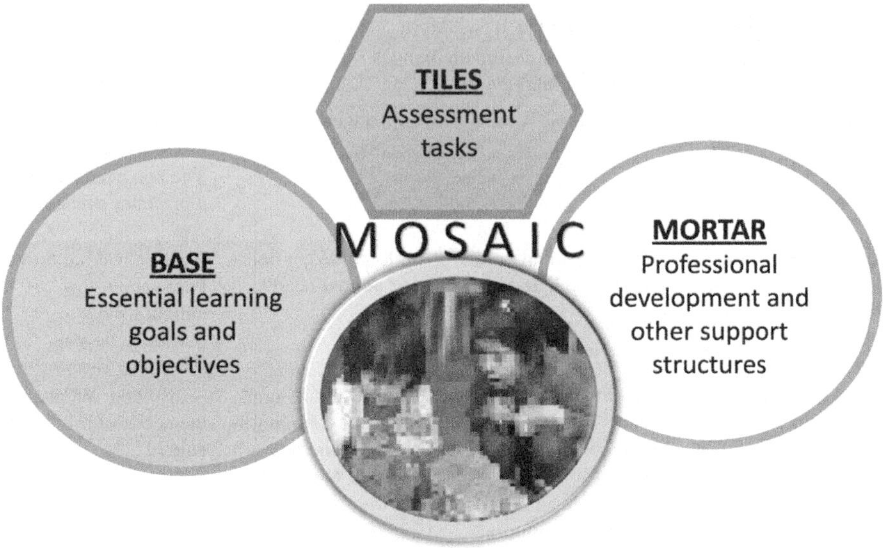

Fig. 7.3 Mosaic Framework

In an engineering design classroom, students demonstrate learning through a variety of means and multiple sources of evidence. Therefore, the Mosaic framework stresses an assessment system where multiple forms of evidence are used to assess student learning and inform instruction. When assessment addresses essential learning objectives and is closely aligned with the curriculum, it can be used for formative and diagnostic purposes that help provide a baseline of student learning to inform instruction and provide a summative measure of student competency and growth. The Mosaic framework entails three key principles to ensure that the instruction and assessment target essential learning objectives, teachers have the abilities to integrate and use multiple forms of assessment, and the tools used to assess student learning are diverse, useful, and appropriate to allow teachers' ability to make meaningful inferences on children's learning.

7.7 Key Principles of the Mosaic Framework

Three critical principles of the Mosaic framework inform assessment of engineering thinking and design competencies (see Table 7.2). These principles may apply to assessment of engineering competencies beyond early childhood education.

Table 7.2 Components of the Mosaic framework

Mosaic framework	Key principles
Base: Essential learning goals and objectives	Targeting essential learning goals and objectives aligned with curriculum, instruction, and assessment
Tiles: Classroom assessment	Using curriculum-embedded classroom assessment tools and tasks that can capture evidence of learning
Mortar: Assessment-centered professional development	Developing teacher competencies to interpret and integrate multiple forms of assessment

7.7.1 Mosaic Principle 1. Targeting Essential Learning Goals and Objectives

Engineering design and integrated STEM curricula often cover an array of learning objectives related to core content areas in science, mathematics, and literacy, as well as engineering practices. Given the rich nature of these curricula and the complexity of the learning environment, the assessment tasks and the scoring guides need to be able to clearly identify and decouple interconnected aspects associated with integrated learning. In Table 7.3, we provide a common set of core engineering design competencies and an associated set of measurable learning objectives for young learners as an example. These competencies target three areas related to: understanding the problem, developing alternative solutions, testing these alternatives, deciding on a final solution, and communicating solution and decisions.

Table 7.3 presents core design competencies for young learners, built on previous work of Douglas, Moore, and Adams' Core Design Competencies for Middle School Students (2016). Douglas et al. (2016) created the competencies for a suite of curricular units for grades 4–8, informed by indicators of high-quality engineering curricula (Moore et al., 2014), and research regarding beginner, informed, and expert engineering design behaviors (Crismond & Adams, 2012). The design competencies are also based on a synthesis of the National Research Council (2012) report, *A ramework for K-12 Science Education,* and the*Next Generation Science Standards* (NGSS Lead States, 2013) for younger elementary grades. In addition, we also considered how others had described design-related competencies and rubrics for design projects in pre-college engineering education (e.g., Groves, Abts, & Goldberg, 2014; Asunda & Hill, 2007). While there is no "one-size fits all" approach to engineering design competencies, we aim to provide educators and curriculum developers a starting point. The purpose of these competencies is to aid educators and curriculum developers in focusing engineering design assessment around what competencies are most important for young children to develop, regardless of the design project context.

Our goal by highlighting a critical set of learning goals and associated learning objectives is to emphasize the engineering skills that can be developed over multiple engineering projects of diverse contexts. A notable characteristic of these engineering

Table 7.3 Core Design Competencies for Young Learners

Design competencies	Design learning objectives
Competency 1: Students engage in problem scoping. Students define the problem, including what is needed and who is affected by the problem.	**Objective 1A**: Students ask questions to define the problem from the perspective of the user. **Objective 1B**: Students provide a rationale for why the problem is important to solve. **Objective 1C**: Students identify the end user's needs. **Objective 1D**: Students describe criteria, i.e., what is needed for an effective solution. **Objective 1E**: Students identify constraints, i.e., the limits placed on a solution to be effective.
Competency 2: Students develop simple models of possible solutions (sketches or physical models). The representations illustrate the functions of the solution. Students use the representations to test and decide which solution to implement and optimize.	**Objective 2A**: Students fluently generate multiple design solutions ideas based on understanding of the problem. Students draw and explain them to allow comparison. **Objective 2B**: Students evaluate various solutions based on criteria identified in problem scoping. **Objective 2C**: Students apply evidence from testing in attempt to improve solution quality.
Competency 3: Students communicate their design solutions through evidence-based reasoning	**Objective 3A**: Students justify their design solution based on how it meets user needs. **Objective 3B**: Students justify their design solution with support from science and/or mathematics concepts. **Objective 3C**: Students articulate the limitations of their design solution.

design competency areas is that they are not specific to a single task but broadly applicable to a series of tasks. This approach of assessing critical learning objectives or student competencies rather than tasks allows tracking learning over time through multiple projects. To illustrate this approach, Table 7.4 represents a mapped set of learning objectives into a set of arbitrary lessons. With a set of core learning objectives that are re-visited with each lesson, such a mapping allows the teacher to focus on these key objectives in his or her observations, when reviewing student work, and eliciting evidence of learning. It is critical to ensure that through instruction, children have opportunities to learn target competencies and that these competencies are what are taught and assessed. We recommend that there are at least three opportunities for children to practice each objective in a semester.

Table 7.4 Decoupling essential learning objectives

Goals	Objectives	Lesson I	Lesson II	Lesson III	Lesson IV	…
Competency 1	1A	x		x		x
	1B		x		x	x
	1C	x		x	x	
	…					
Competency 2	2A	x	x	x		
	2B		x	x	x	
	2C		x	x		x
	…					
Competency 3	3A	x		x		x
	3B		x	x	x	
	3C		x	x		x
	…					

7.7.2 *Mosaic Principle 2. Using Curriculum-Embedded Classroom Assessment Tools and Tasks*

The second principle of the Mosaic framework specifically targets assessment strategies and instruments (tools, tasks, scoring guides, etc.). When identifying, developing, or using any assessment tool, task, or scoring guide, a clear purpose for assessment should be determined. Differentiating the purpose is important as attempts to use an assessment instrument for a purpose different than its intended use can result in misinterpretation of student learning. The assessment instruments must have educational value and a clear alignment with curricular content and instruction. As Chittenden and Jones (1998) argue, especially at early ages, the most critical purpose and function of assessment is to understand student learning.

Beyond rapid developments in cognitive abilities, early childhood is also a time when children are developing their interpersonal competencies. Engineering, being a collaborative profession, helps support collaborative and interpersonal aspects of the learning process. However, the competencies demonstrated through individual tasks and assignments may not reflect competencies developed or demonstrated in group settings. Hence, assessment in an engineering design environment must take into account and recognize the difference between the assessment of cognition related to design competencies and other aspects such as collaboration or care for others.

Simply designing a task without scoring criteria and a feedback guide is not sufficient for assessment to support student learning. We provide sample scoring criteria (i.e., rubric) for providing feedback to students on their design projects (see Table 7.5). An example illustrating how this scoring guide is used can be found in Fig. 7.4 and will be discussed in more detail in Mosaic Principle 3.

Another critical aspect of effective assessment in design lies in the development and use of multiple forms of assessment that provide multiple sources of evidence

Table 7.5 Levels of proficiency for problem scoping, developing and testing models, and evidence-based reasoning

	Learning objective	Advanced	Proficient	Emerging
Competency 1: Students engage in problem scoping. Students ask questions to define the problem, what is needed and who is affected by the problem.	1A. Students gather information to find out more about the problem.	*Student asks three or more questions and collects information to deepen understanding of the problem.*	*Student asks one to two questions to deepen understanding of problem. Little to no additional information gathered.*	*Student did not ask questions to understand problem. There is no evidence of additional information gathering.*
	1B. Students describe the problem and provide a rationale for why it is important to solve.	*Student description of problem includes what is missing/needed and gives a rationale for why this problem is important to solve.*	*Student provides a description of problem including what is missing/needed but without rationale for why it is important.*	*Student description of the problem is unclear.*
	1C. Students identify end user's needs.	*End user is identified and their needs are described in relation to problem.*	*End user is identified, but little to no description of their needs in relation to problem.*	*Unclear end user.*
	1D. Students describe criteria, i.e., what is needed for an effective solution.	*Student described design criteria in alignment with user needs and the problem.*	*Student described most of the design criteria identified are in alignment with user needs and the problem.*	*Student has not identified clear criteria for solution. OR Criteria are unrelated to the user problem and needs.*
	1E. Students identify constraints, i.e., the limits placed on a solution to be effective.	*All explicit constraints have been identified.*	*Some constraints are identified, but 1 or more are missing.*	*Constraints or limits to the solution have not been identified.*
Competency 2: Students develop simple models of possible solutions (sketch, draw or physical models). The representations illustrate the function of the solution. Students decide which solution to implement and optimize.	2A. Students generate multiple design solution ideas based on understanding of the problem.	*Three or more potential solutions are developed with understandable functions. Potential solutions are clearly related to criteria and constraints.*	*Two or three potential solutions are developed which describe function. Necessary criteria are missing or failed to stay within constraintsOR function is not clear.*	*One solution is developed. OR Solutions developed are not in alignment with problem.*
	2B. Students evaluate various solutions based on criteria and constraints (identified in problem scoping).	*Evaluation based on criteria and constraints.*	*Evaluation of potential solutions not fully or explicitly address in alignment with criteria and constraints.*	*Solution is chosen without evaluation.*
	2C. Students apply evidence from testing in attempt to improve solution quality.	*Students performed two or more iterations of testing and making modification for improvement.*	*Potential improvements are identified based on testing.*	*Did not consider improvement of solution based on testing.*

(continued)

Table 7.5 (continued)

	Learning objective	Advanced	Proficient	Emerging
Competency 3: Students communicate their design solution through use of evidence-based reasoning.	3A. Students justify their design solution based on how it meets user needs.	*Justification is provided for final solution, based on how it solves problem.*	*Explanation of final solution has unclear alignment or missing critical aspects identified in problem scoping.*	*No rationale provided for why solution would solve original problem.*
	3B. Students justify their chosen design solution based on connections of science and/or mathematics concepts.	*Explanation of specific features is discussed based on proper use of evidence from testing.*	*Explanation of specific features is discussed based on partial evidence from testing. OR evidence is unclear.*	*No rationale provided for why the final solution is considered effective.*
	3C. Students articulate the limitations of their design solution.	*Recognition of key limitations with relevant reasoning.*	*Recognition of limitations but with incoherent reasoning.*	*No awareness of limitations of a solution.*

on student learning. To obtain such sources of evidence through various assessment activities requires planning around key critical competencies and developing procedures to organize everyday observations into useful evidence. Curriculum and classroom environments should be designed to provide opportunities for students to demonstrate their competencies and dispositions in diverse means.

Effective teachers engage in assessment continuously; yet, people in general consider assessment as a stand-alone activity separate from instruction. Often the terms formative and summative assessment are used to further highlight assessment that is embedded into instruction (as formative) and assessment that is removed away from instruction (as summative). However, we argue that all assessment must have a formative utility although at varying levels. Rather than differentiating assessment types based on the use of the assessment results as formative and summative, we highlight the importance of continuous assessment and introduce terms, opportunistic and structured, referring to the ways we collect assessment data. Various forms of assessments are summarized in Table 7.6; ideally, multiple forms of evidence will be collected to holistically assess learning, and may encompass additional forms not specified here. Additional evidence of learning, beyond the examples shared in Table 7.6, can be compiled in the form of a portfolio.

Portfolios include collections of work that are organized in a systematic way. In such an organized form, the portfolio becomes shareable and useful for not only teachers but also for the child and caregivers. The sharing of the portfolio and discussions around evidence can facilitate partnership between these stakeholders toward the best ways to support student learning. When portfolios are documented at key points in a year, they also become a way to judge a child's progress over time. These can allow, with planning, self-evaluation by the student and reflection. These can also be easily shared with caregivers and allow communication of student competencies and progress without a comparative judgment. Portfolios can also target the assessment of affective components targeting students' attitudes toward learning, content, and the learning environment.

Name: NOAH	**Date:** January 16, 2016

Activity

All students designed living places for the characters they selected in My Little Pony, a popular children's cartoon show.

Noah designed a house of Rainbow Dash and Twilight, two pony characters in My Little Pony.

Observation

1C: Student identifies the end user's needs:

Evidence:

When asked to explain his solution, Noah said, "Rainbow Dash needs a place to jump and play. Twilight likes to study."

2B: Student evaluates various solutions based on criteria and constraints (identified in problem scoping):

Evidence:

When asked to evaluate his solution, Noah said, "This house is good because each pony has her own area, and they can be together over here [far left side]."

When further prompted by: "*Would Twighlight like having the library near the play area?* Noah responded: "*Oh, it may not be quiet for study.*"

Evaluation and Suggestions for Next Developmental Steps:

ADVANCED in 1C: *End user is identified and their needs are described in relation to problem.*

Next Step: Introduce and reinforce the use of terms such as "end user" and "design criteria"

PROFICIENT in 2B: *Evaluation of potential solutions not fully or explicitly address criteria and constraints.*

Next Step: Ask questions to help Noah explicitly list design criteria and then evaluate his solutions based on these criteria.

Fig. 7.4 Example learning chart of student, Noah, which can be used with caregivers. The context of the engineering design challenge is briefly stated. The learning objectives addressed are stated, and then example evidence is described. In the evaluation and suggestion box, suggestions are provided for caregivers and teachers to further support development

Now that we have made a case for collecting evidence in a variety of forms, one would ask how these diverse pieces of information can be integrated to support teacher decision making and ultimately student learning. This can be facilitated by a matrix that shows the alignment between learning goals (curriculum), assessment, and opportunities to learn (instruction). This is simply a mapping of all competencies and learning objectives. Such a matrix can also illustrate areas that are over- or under-assessed. We recommend that curriculum developers present teachers with a set of tools (e.g., pre-designed Excel spreadsheets) to facilitate integration of information. Hence, it is important beyond the assessment tasks and scoring guides to develop tools to compile and organize data so student progress can be visually illustrated.

Table 7.6 Diverse ways of collecting evidence of student learning

	Forms of assessment	Opportunistic	Structured
Evidence of Student Learning	Performance tasks	Observations of student and recordings of any notable observations	Structured activities that involve completion of a task in a specific time frame
	Teacher observations	Anecdotal comments from discussions between teacher and student or discussions among student	Photos of student engagement in a specific task with targeted observations for predetermined competencies and propositions
	Child's work examples	Pictures of artifacts, sketches and drawings, writing when notable progress is observed	Pictures of artifacts, sketches and drawings, writing collected periodically
	Child self-evaluation and reflection	Anecdotal comments from student reflection prompts	Self-evaluation of competencies from a list of predetermined objectives
	Teacher–child conferences	Records of children's talk occuring naturally when a notable comment is made	Records of children's talk facilitated by teacher asking specific questions about the child's work
	Parent–teacher conferences/parent observations	Anecdotal comments from parents	Teacher observations for specific competencies from a list of predetermined objectives

When assessment methods can show student learning over time, they can provide the most useful information informing instructional decisions.

7.7.3 *Mosaic Principle 3. Developing Teacher Competencies in Assessment*

The complexity of engineering design learning settings can make assessment challenging even for an experienced teacher. Teachers themselves are often new to engineering and may not be prepared to identify what aspects of design are important to assess, what level of achievement is developmentally appropriate, or how to use assessment to further support student learning. Hence, any curriculum and assessment development effort must be accompanied with teacher professional development addressing knowledge, skills, tools and strategies for effective assessment that supports student learning.

A recent report, *Transforming the Workforce for Children Birth through Age 8*, outlines foundational knowledge and abilities all professionals who provide direct, regular care and education for young children must have (IOM & NRC, 2015). A notable number of these knowledge and abilities are related to assessment:

1. Knowledge of assessment principles and methods to understand individual children's developmental progression and monitor progress.
2. Knowledge of the principles for assessment necessary to select assessment tools that are developmentally appropriate, unbiased toward specific demographics (e.g., gender, ethnicity, socioeconomic status), relevant, reliable, and valid for chosen purposes.
3. Ability to set appropriate individualized goals and objectives to advance young children's development and learning.
4. Ability to select, implement, and interpret a portfolio of both assessment tools and strategies
5. Ability to use learning trajectories with a deep understanding of the subject, knowledge of the way children think and learn about the subject.
6. Ability to use assessment information to adjust, improve, and individualize instructional practices.

When these knowledge and skills in assessment practices are applied to engineering learning opportunities, there is potential for teachers to further scaffold student learning. The engineering design process itself is iterative with feedback loops to inform decisions along the way. Likewise, high-quality assessment of design practices would allow iterations and multiple opportunities for students to receive feedback and make changes. Professional development could provide opportunities for teachers to practice applying these knowledge and skills to examples of student work and discourse with others.

While assessment and instruction, especially in early childhood settings, need to be tightly melded, it is also important to decouple evidence of learning from the learning activities themselves. A learning activity requires students to complete a task or observe a demonstration or procedure in order to learn. While the same task can be used for assessment purposes, completion of the task is not sufficient. Assessment would then examine the evidence to ascertain the level of learning that took place in the activity, not just that the student performed the activity. Hence, professional development activities should highlight that assessment is a way of collecting evidence to ascertain the extent of learning and what gaps still remain.

Professional development should also explicitly cover how to score and interpret student work with rubrics. Among the most common tools used to evaluate student learning in engineering projects is scoring guides or rubrics because engineering projects are by nature, open-ended. One of the most common mistakes in such situations is attributed to the way scoring guides are designed and used for assessment. While teachers may be familiar with writing rubrics, they are likely to need guidance on the technical aspects. Because the competencies can be assessed over multiple design projects, it is important that they are general enough to be useful in a variety of contexts, but specific enough to actually produce meaningful results. When scoring guides are designed to be either task-specific or hypergeneral, they are not effective in pointing to specific competencies (Popham, 2005). Imagine that a class of students is engaged in several different design projects over the course of a month. Scoring guides that are task-specific focus on aspects of these projects such as the performance of the design prototype on design criteria (e.g., solution can hold 10 rocks without breaking). Hypergeneral scoring guides focus on broad learning goals, that we call competencies (e.g., student communicates ideas well). While task-specific scoring guides would not provide opportunities for educators to see student development between projects over time, hypergeneral scoring guides would not result in useful and specific feedback that can help support student learning and development.

Once a scoring guide that can distinguish different levels of student competencies has been identified, teachers can be provided with explanations for each assessment criteria and examples from students' work and given opportunities to evaluate actual student work. Figure 7.4 illustrates this approach through observation notes about the performance of a specific student (Noah) on two learning objectives (1C and 2B) and explanation of his level of performance. In this assignment, Noah (a six-year old) designed living places for two characters from a children's cartoon show called *My Little Pony: Rainbow Dash and Twilight*. The evidence of learning chart illustrates both advanced and proficient levels of competencies in design. The bottom section of the chart also includes suggestions for further development. In addition, teachers need to be given opportunities to use these tools and integrate them to their daily classroom practices.

It is important to note that often assessment efforts fail when expectations are not evident to children or assessment does not allow useful feedback that can be used for formative purposes. Assessment is a continuous process with many feedback cycles. Effective assessment methods in engineering would serve to not only recognize and judge student design reasoning but also enhance young children's design behaviors to be more informed and systematic. Teacher professional development can cover a variety of strategies to support learning through reflection. One strategy is involving the child in their own assessment (Purzer, Duncan-Wiles, & Strobel, 2013), which can be facilitated through teacher–child conferences and other novel ways such as through narratives or learning stories that capture children's learning (Anthony, McLachlan, & Poh, 2015).

While targeting the key knowledge and abilities outlined above, teacher professional development must also address teachers' predispositions and perceptions about assessment. Assessment has a negative association for some educators, with views that assessment is external and harmful to the teaching and learning process. In fact, research has shown that assessment can hinder *or* support learning, depending on how assessment is conducted and used (Wiliam, 2011). It is important, therefore, to raise educators' awareness that assessment practices supportive of learning are also part of effective pedagogy for engineering.

Classroom assessment is often thought as an interchange been students and educators. However, parents and caregivers play an important role in supporting the development of student learning. In fact, much learning occurs outside the classroom (Stevens, Bransford, & Stevens, 2005) and effective assessment in early childhood education requires the involvement of the student and the caregivers (Grisham-Brown, Hallam, & Brookshire, 2006). Parents can be involved in the evaluation process by sharing student artifacts and anecdotes and by asking to comment on their children's learning and attitudes toward learning at home. Student portfolios are often effective in sharing such information but technological tools (such as educational apps), if available, can facilitate frequent exchange of such information between parents and educators.

7.8 Conclusion

Children are creative and capable of engaging in early engineering when provided genuine design opportunities. Just as it is important to provide high-quality engineering experiences to young children, it is important to support development of engineering thinking through proper assessment. In this chapter, we highlighted the importance of explicitly introducing engineering in early childhood education and discussed prior research related to engineering thinking. In addition, we provided the Mosaic Framework for assessment that supports engineering thinking and learning, as well as presented specific tools and approaches for assessment. For the highest impact on supporting student learning, assessment must be viewed as a multi-faceted coordination like a mosaic that is connected by curriculum design, teacher professional development, and assessment tools and strategies. Such a mosaic should draw a picture of student learning and growth with appropriate learning evidence.

References

Anthony, G., McLachlan, C., & Poh, R. L. F. (2015). Narrative assessment: Making mathematics learning visible in early childhood settings. *Mathematics Education Research Journal, 27*(3), 385–400. https://doi.org/10.1007/s13394-015-0142-2.

Asunda, P. A., & Hill, R. B. (2007). Critical features of engineering design in technology education. *Journal of Industrial Teacher Education, 44*(1). Retrieved from http://scholar.lib.vt.edu/ejournals/JITE/v44n1/asunda.html.

Atman, C., & Adams, R. (2007). Engineering design processes: A comparison of students and expert practitioners. *Journal of Engineering Education, 96*(4), 359–379. https://doi.org/10.1002/j.2168-9830.2007.tb00945.x.

Atman, C., Cardella, M. E., Turns, J., & Adams, R. (2005). A comparison of freshman and senior engineering design processes: An in depth follow-up study. *Design Studies, 26*(4), 325–357. https://doi.org/10.1016/j.destud.2004.09.005.

Chittenden, E., & Jones, J. (February, 1998). *Science assessment in early childhood programs.* Paper Presented at the Forum on Early Childhood, Science, Mathematics, and Technology Education, Washington, DC. Retrieved from https://eric.ed.gov/?id=ED416990.

Crismond, D. P., & Adams, R. S. (2012). The informed design teaching and learning matrix. *Journal of Engineering Education, 101*(4), 738–797. https://doi.org/10.1002/j.2168-9830.2012.tb01127.x.

Daly, S. R., Adams, R. S., & Bodner, G. M. (2012). What does it mean to design? A qualitative investigation of design professionals' experiences. *Journal of Engineering Education, 101*(2), 187–219. https://doi.org/10.1002/j.2168-9830.2012.tb00048.x.

Douglas, K., Moore, T., & Adams, R. (2016). Core engineering design competencies. *Purdue University Research Repository*. https://doi.org/10.4231/R7B56GQZ.

ETA hand2mind. (2013). *Hands-On Standards® STEM in Action Ron's Ramp Adventure*. [Kit]. Vernon Hills, IL: ETA hand2mind.

Gainsburg, J. (2006). The mathematical modeling of structural engineers. *Mathematical Thinking and Learning, 8*(1), 3–36. https://doi.org/10.1207/s15327833mtl0801_2.

Grisham-Brown, J., Hallam, R., & Brookshire, R. (2006). Using authentic assessment to evidence children's progress toward early learning standards. *Early Childhood Education Journal, 34*(1), 45–51. https://doi.org/10.1007/s10643-006-0106-y.

Groves, J., Abts, L., & Goldberg, G. (2014). Using an engineering design process portfolio scoring rubric to structure online high school engineering education. *Practical Assessment, Research & Evaluation, 19*(8), 2. Retrieved from http://pareonline.net/getvn.asp?v=19&n=8.

Hsu, M.-C., Cardella, M. E., & Purzer, Ş. (2010, March). *Elementary students' learning progressions and prior knowledge on engineering design process*. Paper Presented at the Annual Meeting of the National Association for Research in Science Teaching, Philadelphia, PA.

Hsu, M.-C., Cardella, M. E., & Purzer, Ş. (2012). Elementary students' engineering design process knowledge: Instrument development and pilot test. In *Proceedings of the 119th ASEE Annual Conference and Exposition, San Antonio, TX.*

Hunter, L., & Sonter, L. (2011). Wondering why? Exploring scientific thinking in an emergent co-constructed curriculum. *Educating Young Children, 17*(1), 18–19. Retrieved from http://www.ecta.org.au/_dbase_upl/ECTA-EducatingYoungChildren-17.1-2012S.pdf#page=20.

Institute of Medicine (IOM), & National Research Council (NRC). (2015). *Transforming the workforce for children birth through age 8: A unifying foundation*. Washington, DC: The National Academies Press. Retrieved from https://www.nap.edu/read/19401/chapter/21#498.

Jonassen, D., Strobel, J., & Lee, C. B. (2006). Everyday problem solving in engineering: Lessons for engineering educators. *Journal of Engineering Education, 95*(2), 139–151. https://doi.org/10.1002/j.2168-9830.2006.tb00885.x.

Moore, T. J., Glancy, A. W., Tank, K. M., Kersten, J. A., Smith, K. A., & Stohlmann, M. S. (2014). A framework for quality K-12 engineering education: Research and development. *Journal of Pre-College Engineering Education Research, 4*(1), 1–13. https://doi.org/10.7771/2157-9288.1069.

National Academy of Engineering & National Research Council. (2009). *Engineering in K-12 education: Understanding the status and improving the prospects*. Washington, DC: National Academies Press.

National Association for the Education of Young Children. (2012). *NAEYC standards for initial & advanced early childhood professional preparation*. Retrieved from http://www.naeyc.org/files/naeyc/NAEYC%20Initial%20and%20Advanced%20Standards.pdf.

National Research Council. (2000). *Eager to learn: Educating our preschoolers*. Washington, DC: National Academies Press.

National Research Council. (2012). *A framework for K-12 science education: Practices, crosscutting concepts, and core ideas*. Washington, DC: National Academies Press.

NGSS Lead States. (2013). *Next generation science standards: For states, by states*. Washington, DC: National Academies Press.

Novel Engineering. (2016). Retrieved from http://www.novelengineering.org.

Pellegrino, J. W. (2012). Assessment of science learning: Living in interesting times. *Journal of Research in Science Teaching, 49*(6), 831–841.

PictureSTEM. (2016a). *Designing hamster habitats*. Retrieved from http://picturestem.org/picturestem-units/first-grade-hamsters.
PictureSTEM. (2016b). *Designing paper baskets*. Retrieved from http://picturestem.org/picturestem-units/kindergarten-baskets.
Popham, W. J. (2010). *Classroom assessment: What teachers need to know* (6th ed.). NJ Pearson: Upper Saddle River.
Purzer, Ş., Duncan-Wiles, D., & Strobel, J. (2013). Cost or quality? *Science & Children, 50*(5), 34–39.
Purzer, Ş., & Fernandez, T. (2015, October). Teaching and assessment strategies that value innovative thinking. In *Proceedings of the 2015 Frontiers in Education Conference, El Paso, TX*.
Purzer, Ş., Strobel, J., & Cardella, M. E. (2014). *Engineering in pre-college settings: Synthesizing research, policy, and practices*. West Lafayette, IN: Purdue University Press.
Sheppard, S. D., Macatangay, K., Colby, A., & Sullivan, W. M. (2008). *Educating engineers: Designing for the future of the field*. San Francisco, CA: Jossey-Bass.
Stagnitti, K. (2014). Pretend play and literacy. *Educating Young Children, 20*(2), 18–21. Retrieved from http://search.informit.com.au/documentSummary;dn=588774442422469;res=IELHSS.
Stevens, R., Bransford, J., & Stevens, A. (2005). *The LIFE Center's lifelong and lifewide diagram*. Retrieved from: http://life-slc.org.
Wiliam, D. (2011). *Embedded formative assessment*. Bloomington, IN: Solution Tree Press.

Part II
Engineering Curriculum Design and Development

Chapter 8
Engineering Concepts, Practices, and Trajectories for Early Childhood Education

Christine M. Cunningham, Cathy P. Lachapelle and Martha E. Davis

Abstract In this chapter, we examine what age-appropriate engineering should entail for children at the preschool (ages 3–4), kindergarten (ages 5–6), and primary (ages 7–8) grade levels. We propose a set of design parameters that develop foundational engineering concepts and practices in children as they participate in engineering activity and design. At the core, these include understanding engineering as a design process and a focus on materials and their properties. As children engage in engineering, they should work to determine the problem they need to solve, think about criteria for successful designs, and explore which materials are best suited for their needs. They should also conduct tests and reflect upon the results to analyze how well their design worked to solve the problem while meeting criteria. Additionally, engineering education for young learners should foster children's creative thinking, observational skills, and persistence. For each engineering curriculum design parameter, we describe how it can be implemented appropriately for children at each age band (3 and 4, 5 and 6, 7 and 8) based on our experience with children in classrooms.

8.1 Introduction

Engineering is making its way into elementary classroom across the USA and around the world. Curricula like *Engineering is Elementary* and other experimental

C. M. Cunningham (✉) · C. P. Lachapelle · M. E. Davis
Museum of Science, Boston, MA, USA
e-mail: ccunningham@mos.org

C. P. Lachapelle
e-mail: clachapelle@mos.org

M. E. Davis
e-mail: mdavis@mos.org

L. English and T. Moore (eds.), *Early Engineering Learning*, Early Mathematics Learning and Development, https://doi.org/10.1007/978-981-10-8621-2_8

curricula like LEGO Engineering and Novel Engineering have reached millions of children and made important contributions to the field of K-12 engineering education (Cunningham, 2018; Cunningham & Hester, 2007; Rogers & Portsmore, 2004; Wendell, Wright, & Paugh, 2015). Engineering education done well can support children to develop problem-solving skills (National Research Council [NRC], 2012), to develop the engineering and technological literacy that are needed for the twenty-first century (Lachapelle & Cunningham, 2010; Macalalag, Brockway, McKay, & McGrath, 2008; Thompson & Lyons, 2008), and to improve their mathematics and science skills and understanding (Diaz & King, 2007; Kolodner, Gray, & Fasse, 2003; Lachapelle et al., 2011; Penner, Giles, Lehrer, & Schauble, 1997; Sadler, Coyle, & Schwartz, 2000). It can enhance children's engagement and agency in school (Barron et al., 1998; Silk, Schunn, & Cary, 2009) and enhance their motivation for learning, particularly in science, when they see the importance of science in context (Katehi, Pearson, & Feder, 2009; Pearson, 2004; Wicklein, 2006). Introducing children to engineering early on may also increase their later interest in engineering and technical careers, and their enrollment in courses that enable them to pursue such careers (Katehi et al., 2009; Wicklein, 2006).

As the developers of the in-school curriculum *Engineering is Elementary* and out-of-school-time curricula for elementary school, *Engineering Adventures*, and for middle school, *Engineering Everywhere*, we have significant experience in working with and designing engineering experiences for children and their teachers. We have seen that children ages 6–13 benefit from engineering (Lachapelle et al., 2011; Lachapelle & Cunningham, 2010).

Recently, we have turned our attention to the problem of designing for preschool and kindergarten children. These children are capable of engaging in scientific (Jirout & Zimmerman, 2015; Jones, Lake, & Lin, 2008) as well as engineering practices (French & Woodring, 2014) in meaningful ways. Groups like the National Association for the Education of Young Children (NAEYC), the National Science Teachers Association (NSTA), and the National Academies in the United States have all advocated that children engage in age-appropriate practices of disciplines like science and engineering as they learn (NAEYC, 2005; National Research Council [NRC], 2012; NSTA, 2002, 2014).

What should engineering look like with young children? In this chapter, we share our experience and make recommendations. We begin by describing some of the theoretical bases for our curriculum development and lay out a set of curriculum design parameters for engineering that should guide the development of engineering experiences and content for young children. In Sect. 8.2, we offer a series of vignettes at three age bands: ages 3 and 4 (preschool), ages 5 and 6 (kindergarten), and ages 7 and 8 (primary or elementary school) to illustrate how the implementation of each design parameter is adapted for children of different ages and experience, so as to appropriately scaffold their entry into and development of engineering know-how. In Sect. 8.3, we explore the trajectories of these engineering concepts and practices in more detail for each age band.

8.1.1 Theoretical Framework

To guide the development of our elementary engineering curriculum, we base our understanding of learning in the theoretical framework of social constructivism (e.g., Palincsar, 2005). We understand learning to be both actively and socially constructed by the learner and the community (Bransford et al., 2006; Rogoff, 2003; Roth & Lee, 2007). Learning is a developmental process, accruing over time as the learner benefits from experiences with others—especially more knowledgeable others—and the classroom community develops norms and practices that support development (Greeno, 2006; Rogoff, 2003; Roth & Lee, 2007). Particularly in a disciplinary field such as engineering, learners should experience and engage in the practices of the discipline at a developmentally appropriate level (Duschl, 2008; Engle & Conant, 2002; Krajcik & Blumenfeld, 2006; NRC, 2007). Over time, as seen in ethnographic studies of apprenticeship (Collins, 2006; Hutchins, 1995; Lave & Wenger, 1991), learners can move from more scaffolded, peripheral roles in engineering practice—with the curriculum and teacher providing support and guidance—to more independent, central roles (Hmelo-Silver, Duncan, & Chinn, 2007; Rogoff, 2003).

Our theoretical stance on learning is congruent with the positions of leaders and advocates for early childhood education and science learning internationally. The National Science Teachers Association (NSTA) recommends experiential learning and teacher scaffolding, both important aspects of social constructivist learning environments, for the preschool age-group (2014) as well as elementary schoolchildren (2002). The former position statement is endorsed by NAEYC. NAEYC early childhood program standards call for curricula to promote social, physical, emotional, language, and cognitive development, through materials that support teachers to provide a variety of experiences and learning opportunities that meet the needs and interests of the individual children in their programs (2005), as does, for example, the policy statement of the British Association for Early Childhood Education (Stewart, 2011); the Australian Government Department of Education, Employment, and Workplace (2009); and the communication from The Commission on Early Childhood Education and Care of the European Union (2011).

8.1.2 Design Parameters

The curricula that we develop are grounded in this social constructivist learning theory. Our flagship curriculum, *Engineering is Elementary,* is designed for use with children in schools in grades 1–5 (ages 6 through 11). Our out-of-school-time (OST) curriculum *Engineering Adventures* (EA) is developed for children in grades 3–5, and *Engineering Everywhere* (EE) is our OST curriculum for children in grades 6–8 (ages 11 through 14). As part of our engineering curriculum development efforts, we have articulated a set of research-based principles for the design of inclusive curriculum that help us to create materials that attract and engage *all* students, particularly

Table 8.1 Curriculum design parameters overview[a]

Curriculum design parameter	Engineering curriculum and pedagogy should:
1. Narrative context	Set the engineering problem in a real-world, narrative context that is relevant and interesting to children
2. Goals, constraints, and requirements	Explicitly specify a problem to be addressed, as well as constraints and requirements on the solution, in such a way that a variety of valid and creative solutions are possible
3. Exploring materials and methods	Engage children in concrete activities that involve the manipulation of materials and the use of tools
4. Application of science and mathematics	Encourage the purposeful application of science and mathematics content and skills to the design of solutions
5. Analysis of data for planning and redesign	Afford children opportunities to collect data, evaluate their designs, use failure constructively, and reflect on what was learned so they can generate and test new design ideas and solutions
6. Collaboration	Support children to share ideas and materials, to consider each other's ideas, and to negotiate shared solutions
7. Agency	Support children in developing confidence and strategies to solve ill-defined problems
8. Engineering design processes and epistemic practices	Actively engage children in the processes and practices of engineering design while scaffolding their participation

[a]This and all following tables adapted from https://www.eie.org/overview/engineering-trajectories

those that have been historically underrepresented in science and engineering disciplines (Cunningham & Lachapelle, 2014). In addition, we have developed curriculum design parameters that guide the development of social constructivist materials to support children's developing engineering practice (Lachapelle & Cunningham, 2014). These design parameters serve as a common thread in the design of all of our engineering curricula. Our engineering curricula for younger children (ages 3 through 6), currently in development, are also guided by our theoretical stance and our design parameters. In this chapter, we focus on eight design parameters for social constructivist learning in engineering, and how we adapt them to support engineering learning in each age-group (Table 8.1).

The curriculum design parameters will structure our discussion in the following sections. In Sect. 8.2, we introduce three engineering vignettes, one for each of the three age-groups we will discuss, to illustrate the design parameters in action. In Sect. 8.3, for each curriculum design parameter, we describe the research base for the parameter, the research base for its age-appropriate application, and an example of what age-appropriate engineering looks like in classrooms. In Sect. 8.4, we discuss the teacher's role in supporting the engineering development of young children. In Sect. 8.5, we summarize the implications of each design parameter more generally.

8.2 Engineering Vignettes

What does engineering that is consistent with these curriculum design parameters look like at the preschool, kindergarten, and early grade levels? We turn now to offer a vignette at each level. We will use these examples to anchor the discussion of age-appropriate design parameters in the remainder of the chapter.

8.2.1 Engineering for Ages 3–4: Baskets

Preschool engineering needs to be simple and straightforward to accommodate the short attention spans and emergent motor and social skills of 3- and 4-year-old children. Engineering activities can familiarize children with materials and their properties and start to engage them in simple cycles of ask, construct, test, observe, redesign. In the following vignette, preschool children help design a basket that can carry plastic fruit for a salad[1].

> After reading a story about a potluck party, preschoolers in Ms. Jones's room listen attentively as Noodle, the engineering puppet, asks for their help. She describes a problem she is facing and requests all their good thinking to help her solve it. Through a questioning dialogue, children learn that Noodle wants to make a fruit salad for a party, but she needs to get fruit from her garden to the party. Noodle made a basket out of paper to carry fruit, but it isn't working. Through the dialogue, children review with Noodle what a basket is. They see and feel Noodle's broken basket and think about what went wrong with it. Then the youngsters are asked if they can help make a better basket for Noodle.
>
> During the week, children encounter relevant picture books, photographs, basket materials, and tools to help with counting, measuring, and weighing at the Engineering Exploration Center. During Center time, children can use strips of bright yellow plastic, paper, and fabric, as well as clips and a basket frame, to create baskets to carry fruit of a variety of sizes. They test their designs by carrying plastic fruit around the classroom, then counting the number of fruit their basket carried at the Engineering Center. More advanced children can also weigh the fruit and experiment with measuring the "stretchiness" of strips using non-standard linear measures. Children record how many fruit were carried using stickers in their engineering journals, where they either draw or paste a printed photo of their basket, and where Ms. Jones or another teacher scribes or helps them to record what they did and learned. Throughout the week, Ms. Jones also provides thematic enrichment on baskets and fruit in their readings, arts center, science center, and pretend play. At the end of the week, the children gather with Noodle again to tell her about what they did and learned.

[1]This challenge is based on a challenge developed and used by the Discovery Center at the Museum of Science, Boston, in their "Make with Me" space.

Notice how, in this vignette, the curriculum design parameters are implemented in an age-appropriate manner. Context setting (Parameter 1: Narrative Context) and problem definition (Parameter 2: Goals, Constraints, and Requirements) involve an interactive puppet and physical (nonfunctional) example of a common technology the children are challenged to help improve. The function of the technology itself is aligned to children's experience and vocabulary: It is a tool for carrying. Stories, pictures, and other physical examples are provided, as well as opportunities and tools for materials exploration (3: Materials and Methods). Testing is easily accomplished by children through counting and weighing (4: Application of Science and Mathematics) and evaluated by observation (5: Analysis of Data for Planning and Redesign). Neither reading nor writing is required. Collaboration is not required: Children work together or alone at will, with the support of the teacher to share materials and ideas (6: Collaboration). Children engage directly with materials to test their own ideas, with support from the teacher as needed (7: Agency). Children engage in processes of questioning, investigating, designing, evaluating, reflecting, and communicating with significant support from adults (8: Engineering Design Processes and Practices).

8.2.2 Engineering for Ages 5–6: Sails

Kindergarteners, too, need engineering experiences that are designed with their capabilities in mind. The following vignette illustrates the design parameters applied to an engineering activity for kindergartners, in which children design sails for a Styrofoam sailboat. If the boat catches the wind from a fan, it will glide on straws down a fishing-line track.[2]

> Kindergarteners in Ms. Martinez's class are excited—they are going to get to engineer their own sails! The "Ask" step of their engineering unit begins with a short story featuring Janelle and Malik, who design sails to carry messages across a 12-foot gap between their apartment windows. The afternoon is windy, and Ms. Martinez takes her charges outside to explore how wind has energy that they can catch and use. The children fly a kite, observe the flag flapping, watch a pinwheel spin, and stand so they can feel the wind pushing on their backs and rushing into their faces.
>
> Back in the classroom, the children talk about sailboats, look at some pictures of boats with sails of different shapes, and watch a short video of sailboats moving. Children discuss what a sail must do to make a boat move (capture the wind). They learn more about their challenge—to engineer a sail for a "boat" that will glide as far as possible down a track. They are introduced to the materials they can use to create it: index cards, copy paper, wax paper, tin foil, and plastic bags. As a class, they create a "materials and their properties" table and list words that describe each material.
>
> Then, children work side by side at their tables, sharing materials and ideas as they each sketch and create sails to test on the sail track. Ms. Martinez prompts them to observe what works well and what doesn't, and the children redesign and retest many times, counting how may tiles on the floor each sailboat travels. At the end of the day, each child sketches a "final"

[2]This activity is a modification of a lesson in the *Engineering is Elementary* "Catching the Wind: Designing Windmills" unit.

sail design and labels the materials. They gather on the rug for a "Shareout," where children display their sails and sketches, describe them, and talk about what they tried and learned. The reflection discussion, led by Ms. Martinez, surfaces that sail size and sail material are important parameters for sail performance. Tomorrow they can use what they learned to try again if they choose during a "free play" session.

The differences between the preschool vignette and the kindergarten vignette are subtle but important. Setting the Narrative Context (Parameter 1) in preschool involves a simpler story conveyed directly to accommodate very short attention spans, while the kindergarten experience can involve more reading and discussion. Likewise, exploring the slightly more complex Goals, Constraints, and Requirements (Parameter 2) as children define the scope of the problem can involve more variables, more activities, and more multimedia examples. The Engineering Design Processes and Practices (Parameter 8) are again supported and explicitly discussed by the teacher, but with more extensive engagement in and discussion of the practices of engineering; the same is true of children's explorations of Materials and Methods (Parameter 3). The teacher supports children to think about their designs in the context of the science of air movement and energy, to collect data about their designs through nonstandard measurement, and to discuss sail size in relation to sail success (4: Application of Science and Mathematics). Children can easily judge how well a sail works by observing how far each boat moves down the track, which the teacher leverages to help children compare their sail designs and derive some basic principles they can use to improve (Parameter 5: Analysis of Data for Planning and Redesign). Collaboration (Parameter 6) is more formal—children are assigned to groups or pairs to share materials and are encouraged to discuss their ideas, but usually each child will work on his or her own design. Children work more independently in kindergarten than in preschool, still with the Agency (Parameter 7) to determine their own solutions.

8.2.3 Engineering for Ages 7–8: Hand Pollinators

The *Engineering is Elementary* "The Best of Bugs: Designing Hand Pollinators" unit provides an example of how the curriculum design parameters are applied for children who are 7 or 8 years old. In this unit, children meet "Mariana," a young engineer in a story. Mariana has a problem to solve: A plant in her garden frequently flowers, but never produces any berries. Through observation, she learns that insect pollinators are not visiting her plant. She needs to pollinate the flowers, but how?

Today, first graders in Ms. Chen's class begin their work as agricultural engineers. Ms. Chen has read the story of Mariana to the class; they discussed how Mariana solved her problem by making a tool that moved pollen from one flower to another by hand. Now, the children in the class are challenged to engineer their own hand pollinators using a five-step engineering design process: Ask, Imagine, Plan, Create, and Improve.

Children begin by reviewing what they know about insect pollination. They discuss bees, butterflies, and the parts of flowers. They identify the important features of a hand pollinator—that it picks up and drops off pollen, like the legs of a bee. With a partner, each group investigates the materials they will have available to use for their hand pollinators. Ms. Chen gives them lots of different choices—marbles, erasers, tin foil, tape, pom-poms, and pipe cleaners. The class identifies properties of each: Is it fuzzy, shiny, pink, etc.? The children test each material by dipping it in baking soda then trying to tap the baking soda off. Ms. Chen guides their discussion of the properties of each material and the group reflection of how well each material works to pick up and drop off pollen.

Children then brainstorm ideas for their pollinator designs. They plan out their initial ideas. "This is how we are going to make ours," says one child as he proudly holds up a sketch of a pom-pom attached to a twisted pipe cleaner. Each part of his sketch is carefully drawn and labeled. Finally, pairs of children work together to construct their pollinator and test it to determine whether and how well it picks up and drops off pollen. Their observations fuel new ideas and they return to improve their first design.

In this vignette, the storybook and class discussion provide the Narrative Context (Parameter 1) and draw out children's prior knowledge and experience, including science knowledge of insects, plants, and the pollination process (Parameter 4: Application of Science and Mathematics). The children have time and support to define the scope of the problem (Parameter 2: Goals, Constraints, and Requirements) and to explore the Materials and Methods (Parameter 3). The teacher explicitly reviews steps in a simplified Engineering Design Process, and Engineering Practices are supported by written materials, as well as the teacher (Parameter 8). Both simple, written materials and the teacher's support of class discussion are used to scaffold children in the Application of Science and Mathematics (Parameter 4) and Analysis of Data for Planning and Redesign (Parameter 5). Children judge how well their designs succeed using a standard testing procedure, with the support of the teacher, who helps them to communicate their ideas and findings (Parameters 5 and 8). Collaboration (Parameter 6) is in pairs or groups of three on a shared design solution, with support from the teacher as needed to negotiate the shape of that shared solution. The teacher may model how to create solutions, but children have the Agency (Parameter 7) to determine their solutions for themselves.

8.2.4 Development Across the Age Span

Activities for all age levels apply social constructivist principles of learning and pedagogy in a manner appropriate for the developmental level of the children in that age-group. Children need the opportunity to construct their own understanding (Piaget, Gruber, & Vonèche, 1977). Children also need the social support to accomplish with the help of others what they are not quite capable of accomplishing themselves—support to learn in what Vygotsky named the *zone of proximal development* (Vygotsky, 1978). Of course, some children will be ahead of the typical developmental trajectory in some ways, and others behind, so the teacher will need to use professional judgment to flexibly meet the needs of all children.

The three vignettes given in Sects. 8.2.1 through 8.2.3 illustrate differences in curriculum and pedagogy that reflect differences in child development across this age range. Between age 3 and age 8, children acquire considerable experience with the world, as well as experience with learning from books and other media. They are able to tackle greater and more complex challenges as they grow. Their skills in communication, mathematics, science, and literacy develop dramatically. Vocabularies increase tenfold between age 3 and kindergarten (Farkas & Beron, 2004). Children become readers and writers in early elementary school. Children also develop socially: At age 3, they are typically still having trouble with sharing and cooperating, but by age 8 they have become accomplished at turn-taking in games and at school (Copple & Bredekamp, 2009). Emotionally, children develop in their executive functioning skills, ability to plan thoughtfully, ability to cope with frustration, and ability to deal with failure (Allen & Kelly, 2015). Physically, their motor skills also develop in this time: 3-year-olds typically produce scribbles and perhaps a few letters, often written backwards; by age 8, children are able to write well-formed sentences on wide-lined paper and draw recognizable sketches (Copple & Bredekamp, 2009). Perhaps most importantly, children become more independent between ages 3 and 8, developing the ability to follow directions and maintain attention over longer periods (Copple & Bredekamp, 2009).

8.3 Developmental Trajectories for Engineering Curriculum Design

Using the vignettes, let us look closely at some important features of each engineering experience and think more deeply about development across these age span, and how those changes should be reflected in well-designed engineering activities. The curriculum design parameters we outlined in Sect. 8.1 structure this section.

8.3.1 Narrative Context

Setting the context is always important in developing an engineering problem for children. An appropriate context helps children to understand the problem at hand and its relevance (Buxton, 2010; Klassen, 2007; Kolodner, 2006). The context helps children to see why engineering is important and to understand its role in the world (Brophy, Klein, Portsmore, & Rogers, 2008). Setting learning in a narrative context helps children to understand that engineering is done for a purpose: to meet the needs of people or the environment (Brophy et al., 2008; National Research Council [NRC], 2011).

Narrative is a particularly powerful—and human—vehicle for children's learning (Wilson, 2002). It is a way of sharing information with children that is accessible and that makes human connections. It helps children to develop an understanding of types of situations where their learning may be relevant, affording transfer to new situations (Kolodner, 2006).

Table 8.2 Developmental trajectory for design parameter 1: narrative context

What it looks like: ages 3–4	What it looks like: ages 5–6	What it looks like: ages 7–8
• The teacher presents the context through dialogue and role play, e.g., with a puppet • The teacher scaffolds children's dialogue • The topic is within children's experience, but still requires discussion and support • The teacher reads supplementary stories, provides pictures and exemplars, and supervises other supplementary experiences to expand children's knowledge base	• The context can be presented through characters in a short picture book • The teacher reads aloud and supports comprehension through questioning • The topic involves a child's personal problem, familiar to children of this age • The teacher reads supplementary stories, provides pictures and exemplars, and supervises other supplementary experiences to expand children's knowledge base	• The context can be presented through characters in a longer picture book • The teacher reads aloud and supports comprehension through questioning • The topic is familiar to children indirectly through texts and media • The teacher reads fiction and nonfiction books, provides video clips and exemplars, and supervises other supplementary experiences to expand children's knowledge base

Narrative is also a powerful way to engage and motivate children. Emotional engagement leads to deep, transferable learning (Immordino-Yang, 2015). Children are motivated to learn when they understand the reason for what they are learning and when they see that reason as important to them (Buxton, 2010; Klassen, 2007). Stories where people are making a difference in the world are particularly engaging for children (Brotman & Moore, 2008). It is especially important to develop engineering stories and challenges that interest girls and racial minorities (Katehi et al., 2009). Too often, engineering for children focuses on structures, vehicles, and gadgets, which are more interesting to boys, and can reinforce gender stereotypes about engineering (Clark & Andrews, 2010; Miller, Blessing, & Schwartz, 2006).

Presenting an engineering problem or challenge within a narrative context also allows for integration of engineering with literacy instruction—a strong advantage in the education of young children, for whom the development of literacy is a primary goal. As children gain experience and knowledge about the world, and learn to read and to learn from text, the media through which the context is delivered can change. The context can be real or fictitious and can be introduced through a variety of narrative media—puppets, stories, letters, videos, interviews. However, the story should focus on relatable characters that engage children's imagination. EiE's formal and informal curricula use all of these media to engage children and set the stage for the engineering unit (Table 8.2).

8.3.1.1 Ages 3 and 4

Preschool children, who have comparatively little experience with the world, need support in their understanding of a situation where a problem exists to be solved. A "script" is a mental representation of a type of event with key features specified—for example, people bring gifts to a birthday party, wear nice clothes, and sing to the guest of honor—that children develop to help them understand new experiences (Kolodner, 2006). The use of a picture book to introduce the context—in Ms. Jones' class, the book introduces a potluck party where the guests bring food—gives children a script for understanding the situation described by Noodle and the need or want (to carry fruit to the party) that motivates the engineering challenge.

The youngest children respond best when they can engage directly with the design challenge. Role play is a highly motivating format which supports children's social and emotional development, and preschool children love to engage in role play with adults (Elias & Berk, 2002; Vygotsky, 1978). The use of a puppet allows children to engage directly with their "client" in a fun and playful game of pretend.

8.3.1.2 Ages 5 and 6

By ages 5 and 6, children have accumulated a few more years of experience, but much of the world is still new to them. Again, it is important to develop the context of the situation motivating the design challenge. In Ms. Martinez's class, this is accomplished with the story, where children learn about other children their age who faced a similar challenge—sending messages across a gap—and decided to use their fans and "sailboats" to engineer a solution. This challenge engages children's imaginations, as children identify with the story and characters (Stinner, 1996).

As with preschool, the teacher scaffolds the background development and research process, both by providing supplementary materials (pinwheels, videos of sailboats, etc.) and experiences (playing with pinwheels, taking a walk on a windy day) to enrich children's knowledge of the natural phenomenon of wind and of technologies that make use of it. She scaffolds the questioning process as well, helping to surface what children are wondering about and helping them to express their ideas as questions to be investigated.

8.3.1.3 Ages 7 and 8

By age 7, children are more familiar with narratives containing information, and improved reading comprehension of such texts is a curricular goal for this age band (e.g., Australian Government Department of Education, Employment, and Workplace, 2009; Common Core State Standards Initiative, 2012a; Department for Education and Skills, 2006). It makes sense to incorporate the context setting into reading time in the classroom, especially as science time tends to be more limited. The teacher can use standard reading comprehension pedagogy to help children understand the motivation for the design challenge, as well as relevant science and engineering information. Just as with the younger children, characters and their dilemmas provide children with an engaging entry to the problem, and a "story to think with"—a narrative that helps them make sense of aspects of the world (Wilson, 2002).

8.3.2 Goals, Constraints, and Requirements

Before beginning the process of engineering design, engineers work to define a given problem and its scope, sometimes called "problem scoping." The problem, as well as constraints on and requirements for the solution, is almost always defined by a client (Brophy et al., 2008; Cunningham & Carlsen, 2014). When children take on a "client's" project, they learn about how engineers function in the real world. They also have the opportunity to learn to take on the client's perspective, to think about what would make a good or preferred solution for the client (Brophy et al., 2008).

Building in constraints on and requirements for the design challenge has benefits for learning as well. By focusing on a constrained design challenge, it becomes much more straightforward to build learning objectives in science, math, and social studies. Children can focus on the learning objectives and measure their own progress, both in meeting the design challenge and in learning (Kolodner, 2002; Sawyer, 2006a).

Constraining the design challenge does not mean that creativity must be constrained. On the contrary, both discipline and creativity are important to engineering (Stouffer, Russell, & Oliva, 2004). Innovation requires balancing constraints and requirements while maximizing outcomes for complex situations, and children can learn a great deal about "disciplined improvisation" from working together on challenges that are both open-ended and constrained (Sawyer, 2006a).

Children need to learn to define the scope of the problem: the goal and requirements for the solution, as well as constraints on solving it. The specific engineering problem that children are asked to define and solve must be accessible and relevant to the given age-group. It must be something with which they have enough experience to understand the scope and envision a solution. The problem should be compelling enough that children care to solve it. Developmentally, this will look different for children who are 3 than for 8-year-olds. Older children can handle more constraints and complexity than younger children (Table 8.3).

8.3.2.1 Ages 3 and 4

The youngest children, those in preschool, should be challenged to create designs that are concrete and physical—well within their experience—so they can focus on design, rather than on trying to understand the function of the object (Bairaktarova, Evangelou, Bagiati, & Dobbs-Oates, 2012). Design challenges should be simple, with few steps. For example, 3- and 4-year-olds in Ms. Jones's class are asked to make a basket to carry fruit. Even if some children have never seen or played with a basket, they understand its function—to carry things. Becoming familiar with baskets is easy to accomplish with physical examples and pictures. Children also benefit from

Table 8.3 Developmental trajectory for design parameter 2: goals, constraints, and requirements

What it looks like: ages 3–4	What it looks like: ages 5–6	What it looks like: ages 7–8
• The technology is something children have used before • Children design a technology that is functional and affords imaginative play • No more than two criteria for success are specified • A small assortment of available materials affords a variety of valid solutions while not overwhelming children	• The technology is something children have seen or heard about • Children design a technology or model with a function that can be understood with some hands-on experimentation • Up to three criteria for success are specified • A variety of available materials and methods afford a variety of valid solutions	• The technology may be new to children • Children design a technology or model with one or two functions that are readily understood with instruction • Up to four criteria for success require trade-offs • Balanced trade-offs ensure that many valid solutions are possible

additional enrichment activities and materials to help increase their knowledge base and broaden their experience of baskets and other technologies that are used to carry things. Books, pictures of baskets, and sample baskets all serve this role in Ms. Jones' class.

To be developmentally appropriate for this age-group, the goal should be for children to create something real or a model of something familiar, like a house made of blocks, to readily engage children's imaginations, because role play is a dominant form of learning for this group (Henricks, 2008). The technology they design should be functional to afford role play and hands-on exploration. In our example, the basket is used to collect and carry ingredients so that Noodle can make a fruit salad.

The materials and methods available for children to solve the problem are more limited than they will be for older children. Strips for the basket are pre-cut, of no more than 10 or 12 materials. Only one or two similar colors are represented, so children can focus on more relevant attributes, such as durability and stretch. A handful of types of fasteners are provided. These constraints reduce the complexity for small children to something manageable for them to explore and test. Challenges and activities should accommodate children with a range of motor skills and tolerance for frustration.

Children of this age need plenty of support for asking questions about and defining the scope of the challenge: both to frame proper questions and to engage in the questioning process (NSTA, 2014). Preschool teachers can engage children in this practice by modeling how to ask good questions and by restating children's questions more clearly for the group (Epstein, 2006; French & Woodring, 2014; Jirout & Zimmerman, 2015).

Our vignette for this example illustrates how the teacher supports children to define the scope of the problem.

With the help of Ms. Jones, the children learn that Noodle's fruit-carrying basket is broken. She passes around the broken basket so the children can look at it. Then, with Ms. Jones's prompting, the children ask Noodle further questions about the problem. They observe how the paper the basket is made from is ripped, and some strips have pulled free from their fastenings. When Noodle asks the children to help her engineer a better basket from stronger materials, they are enthusiastic.

8.3.2.2 Ages 5 and 6

In kindergarten, most children's attention spans allow them to begin problem scoping by discussing the goal of the design challenge as a class. Ms. Martinez's class begins by revisiting the story and talking about how the class design challenge will be similar to and different from the one featured in the story.

Children who are 5 and 6 years old also need significant teacher support to define the scope of the problem; however, they are able to manage more constraints or requirements. Younger children need to focus on only one or two dimensions of the problem and a tightly constrained selection of materials—whether the strawberries stay in the basket and how many strawberries fit, with a limited number and color of pre-cut types of strips to choose from. Conversely, children aged 5 and 6 can focus on two or three dimensions, with more room for variation in materials and methods for solving the problem—in this case, how far the boat travels and whether it stays upright on the string are the criteria for judging success. The constraints for children in this age range are looser, with more materials to choose from, and children are free to cut the materials on their own or combine them.

At this age, children are ready to work on technologies that are more novel to their lived experience. Ms. Martinez's class is designing sails. Few children of this age will have seen a sail up close, let alone played with it, but they know or can quickly learn its common forms. More importantly, its function—to capture wind and put that energy to use to move a boat—is one they readily grasp with a little hands-on experience.

In Ms. Martinez's class, Aliya's first idea is to use white fabric for her sail. She carefully cuts the fabric into a triangle. "See, it's just like the boats in the movie," she tells Malcolm, showing him the sail she carefully taped to the popsicle-stick mast.

The important feature is ensuring that children have a place to start—that they know enough to begin to solve the problem. Later, children will have the opportunity to realize that the color does not matter, and other shapes besides a triangle may work better for catching wind with a boat on a track. They can experiment with methods such as building frames from popsicle sticks or pipe cleaners, and they can experiment with rigid materials such as card stock as compared to flexible materials like sheets of thin plastic or fabric.

8.3.2.3 Ages 7 and 8

Seven- and eight-year-old children can spend more time defining and exploring the scope of the problem. The teacher will still set out the challenge with the whole class and support children in thinking about it, but children will also spend time exploring and experimenting with materials in their small groups. They can go beyond "messing around" and simple qualitative evaluations to conducting "fair tests," collecting data, and discussing the implications. Materials and methods for combining, altering, or using them can have more variety and challenge, both because children are better able to focus on important properties and because their dexterity is improved.

Children of this age are ready for more complex design challenges. While still hands-on and experiential, a challenge can be more abstract, such as designing a process or a model that represents something they are just learning about. The technology children are challenged to build can be something that most have never seen or thought about, such as a hand pollinator for flowers. In this case, children can be introduced to the technology and need for it through a story, video, picture, or conversation. However, the technology still should function in a predictable, straightforward way and be easy to evaluate.

8.3.3 Exploring Materials and Methods

Exploration of materials is an important part of defining the scope of the problem, where children gather information and explore the dimensions of the problem so they can better understand how they might solve it (Atman et al., 2008). It is an extension of the exploration of the context, criteria for success, constraints, and background resources that inform children's understanding of the design problem in the larger sense.

By exploring materials and methods, children construct a more specific understanding of how they can make use of the materials and tools provided (Brophy et al., 2008; Cunningham, 2018; Roth, 2001). In any given classroom, some children will have spent more time using tools such as scissors, rulers, tape, glue, paper clips, and screwdrivers than others. Some children will have had more experience than others playing with dirt, sand, and rocks; paper, cardboard, and cardstock; or felt, fabric, and ribbon, for example. In any classroom, allowing children sufficient time to become familiar with the physical properties of materials and proper uses of tools enables the more experienced children to teach their peers and gives less experienced children a chance to catch up. It also prepares children for later work with models, simulations, and other abstractions from the physical.

In developing our units, we are careful to pilot test materials and methods with children from a range of backgrounds to ensure that the variety presented is sufficient to interest children but not to overwhelm or distract them from learning about key properties. Methods of manipulating and combining materials, and the tools children use, should be designed to contribute to learning objectives, not to lead children

Table 8.4 Developmental trajectory for design parameter 3: exploring materials and methods

What it looks like: ages 3–4	What it looks like: ages 5–6	What it looks like: ages 7–8
• Children explore a single physical property of all materials and group or arrange them according to that property with significant teacher support • Children explore a basic method or tool (e.g., taping, cutting, fastening with elastics)	• Children explore two properties of the materials that are important to the design and compare materials for suitability with significant teacher support • Children make use of a variety of basic tools and methods for construction	• Children explore, describe, compare, and evaluate the properties of materials for use in a design solution with teacher support • Children make use of a variety of methods and basic tools for construction, including specialized methods (e.g., folding paper to create a beam)

on long tangential explorations. Though such side projects may be interesting to children and productive in some ways, we believe that it is important that the designed curriculum be as focused as possible, and we leave decisions about pursuing other interests and extensions to the teacher (Table 8.4).

8.3.3.1 Ages 3 and 4

Children as young as 3 and 4 benefit greatly from the opportunity to examine a physical example of a problem, because they have so little experience to draw on for understanding it (Roth, 2001). Similarly, children benefit from the opportunity to directly manipulate the materials available for creating a design. With physical manipulation, children can construct a more comprehensive understanding of the problem, including sensory-motor experience they can draw on as they work to make sense of the world (Brophy et al., 2008; Copple & Bredekamp, 2009; French & Woodring, 2014). In Ms. Jones' class, children acquire valuable experience as they manipulate the broken basket and the materials available for creating a new basket. These physical interactions can serve as a launching point for understanding the properties and limitations of materials, the principles of basket construction, and the weaknesses of the given basket design. However, construction should be easy for small hands with emergent motor skills to accomplish. Rather than using tape or glue, children use clips such as hair clips and clothespins to attach strips of fabric, plastic, or paper to the basket frame.

Making observations of materials and communicating what they see are scientific and engineering practices that children need support to enact; teachers play a vital role as they scaffold children to increased competence within the zone of proximal development (Copple & Bredekamp, 2009; Jirout & Zimmerman, 2015; NSTA, 2014). Ms. Jones supports her children to make detailed observations and communicate what they see as they examine the broken basket and materials available for the challenge.

Vocabulary development is another key facet of learning at this age (Copple & Bredekamp, 2009). Preschool children learn new vocabulary at a prodigious pace given rich environmental input (Farkas & Beron, 2004). Ms. Jones asks children to describe the materials and also provides new vocabulary as the children need it, helping them to express their observations with more precise language and expand their reservoir of descriptive words.

In this vignette, note how Ms. Jones works with the children to scaffold their sensory experiences into more sophisticated observations and analysis of the problem than they would have been able to accomplish on their own.

> Ms. Jones passes around strips of fabric, plastic, and paper, as well as a variety of clips, and encourages children to play with them and make observations about their properties, as she supports their learning and use of relevant vocabulary (stretchy, strong, etc.). Later that day, 3-year-old Jackson sings while he works at the Engineering Exploration Center with colorful strips of fabric, carefully attaching them to the frame with large clips. "Stretchy, stretchy, stretchy," he chants softly to himself. When Ms. Jones stops by to see how he's doing, he proudly shows off his creation. "I can carry so many strawberries! We're going to make lots of fruit salad," he tells her.

8.3.3.2 Ages 5 and 6

As in preschool, children benefit from hands-on manipulation of materials, scaffolding of observation practices, and vocabulary development (French & Woodring, 2014; NSTA, 2002). Again, because of the longer attention span, children can sit for a longer period as they help the teacher document their observations in a chart (Fig. 8.1).

> The children sit on a rug as Ms. Martinez passes out samples to of the materials they can choose from when designing their sails. She gestures to the easel, where she has prepared chart paper with two columns and a number of rows (see Fig. 8.1). The first column is labeled "Material"—a sample of each material is taped in its row. The second column is labeled "Properties." Ms. Martinez calls on children to name properties for each material. For the felt sheet, Tammy says, "Bendable." "How do you know?" Ms. Martinez asks. "I can bend it and touch the ends together," Tammy replies. Jared calls out, "It soaks up water too." "Another word for that is 'absorbent,'" Ms. Martinez tells him. Next to the sample of the felt, Ms. Martinez writes "bendable" and "absorbent."

8.3.3.3 Ages 7 and 8

Children of this age band benefit not only from hands-on exploration, but also from teacher scaffolding of their observations, helping children to develop their vocabulary about materials (French & Woodring, 2014; NSTA, 2002). Children have the fine motor control to use scissors, folding techniques, and tape and are learning how to properly use rulers, basic thermometers, liquid measures, and other standard measurement tools.

Curriculum for grades 1–2 should approach problem scoping in a similar way to kindergarten, but with more depth of content and discussion. In addition to discussing

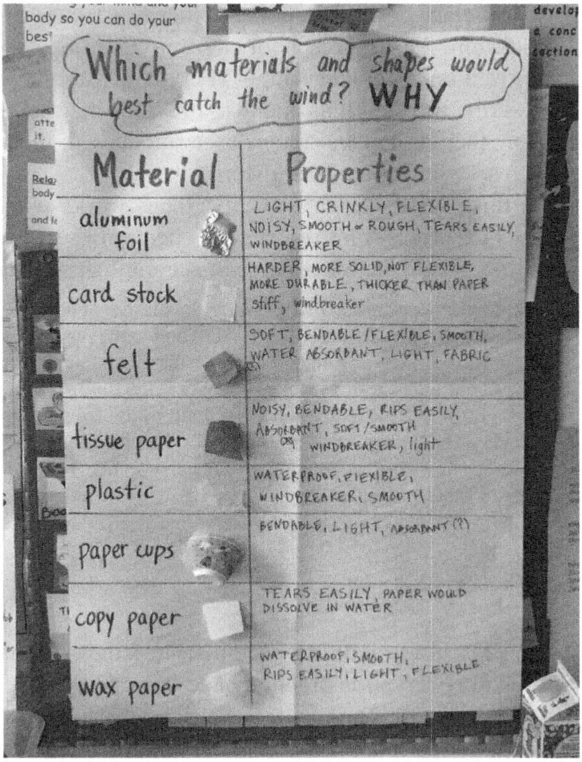

Fig. 8.1 Materials and their properties chart

the goal of the design challenge, observing properties of samples of materials, and adding to their vocabulary for describing materials, children are ready to discuss which properties are most important for their design and to make justified predictions about which materials will perform the best (NGSS Lead States, 2013). Children of this age are also ready to systematically test materials for suitability for use in their design. As in Ms. Martinez's class, the children create a Materials and Properties table. However, when this is completed, Ms. Chen asks her students to think about what the pollinator must do and make predictions about which of the properties are the most important.

> "We're going to test and compare some different materials today to see which are good choices to use in a hand pollinator," Ms. Chen informs the class. "What must a good hand pollinator be able to do?" Maria raises her hand. "It needs to pick up pollen," she says. Her partner Daniel chimes in, "It needs to drop off the pollen too, on the next flower." Ms. Chen praises their thinking. She then introduces the children to the materials they will test (e.g., pom-poms, marbles, etc.) and collects their contributions. She circles the properties children predict will be most important for their hand pollinator designs. The children will return to reflect on their predictions after testing.

Table 8.5 Developmental trajectory for design parameter 4: application of science and mathematics

What it looks like: ages 3–4	What it looks like: ages 5–6	What it looks like: ages 7–8
• The design challenge connects with basic life, Earth, or physical science principles • Children practice counting and using qualitative measures	• The design challenge connects with learning objectives from age-appropriate science content • Children use nonstandard measures, use simple arithmetic, and collect simple data with structured support	• The most successful design solutions will take scientific considerations into account from age-appropriate science content • Children use standard measures, calculate scores, and collect and record data

8.3.4 Application of Science and Mathematics

Science and mathematics are vital to the practice of professional engineering. Conveying the relationships between these disciplines helps children to better understand their importance. Integrating them meaningfully can improve children's learning and achievement across the subjects (Brophy et al., 2008; Katehi et al., 2009; Kolodner, 2002; Lachapelle et al., 2011; Oh, Lachapelle, Shams, Hertel, & Cunningham, 2016; Roth, 2001; Zubrowski, 2002), including engineering (Lewis, 2005).

Young children are capable of conceptual learning in science, as well as scaffolded engagement in practices such as asking questions, predicting, observing, and explaining (Jirout & Zimmerman, 2015; NRC, 2007). Science can be productively integrated with engineering projects (Kolodner, 2002; Kolodner et al., 2004). At all grade levels, engineering activities offer a rich opportunity and compelling context for students to apply what they understand about science to a design problem (Brophy et al., 2008; Katehi et al., 2009; NRC, 2012; Zubrowski, 2002).

In mathematics in the USA, the NAEYC and National Council of Teachers of Mathematics (NCTM) have jointly issued a statement delineating appropriate conceptual content and progressions through age 6 (2010). In Britain, the Department for Education and Skills has made similar calls (2006). Engineering can provide another avenue for learning in these areas, adding to the variety of opportunities children need to develop understanding (Barron et al., 1998; NAEYC & NCTM, 2010) (Table 8.5).

8.3.4.1 Ages 3 and 4

In preschool, science content most appropriately engages children in investigations of the world around them, both indoors and out, through play and exploration (Department for Education, 2014; NSTA, 2014). Young children are capable of engaging in a developmentally appropriate way with scientific practices, with the support of adults (Jirout & Zimmerman, 2015). Adults play a central role in scaffolding children's skills and understanding within the zone of proximal development, to gradually build children's capacities over time (Jirout & Zimmerman, 2015). Adults also have an important role in structuring children's environment to support learning through play and direct experience, as well as providing opportunities for cross-disciplinary learning in math, literacy, engineering, science, and other disciplines through extended projects (Copple & Bredekamp, 2009; NSTA, 2014).

> After all the children have had time to pass around and play with the materials, Ms. Jones holds up a strip of plastic. "Tell me what you observe about this material," she says. "When I do this it doesn't move," says Emma, pulling at both ends. "It's not very stretchy, is it?" asks Ms. Jones. "Not stretchy," Emma agrees. "It's shiny," offers Jackson.

Preschool mathematics content includes counting skills: Specifically, children learn that counting involves a one-to-one correspondence of number names to items and that number names progress in a defined order, neither skipping nor repeating. Three- and four-year-olds are also expected to practice using nonstandard measures to judge quantities as more, the same, or less in preschool (NAEYC & NCTM, 2010). In Ms. Jones's room, both of these mathematical practices are put to use in context.

> Jackson carries his basket, stuffed with plastic strawberries, to the counting station. "How many strawberries fit in your basket?" Ms. Jones asks, and she helps Jackson to count as he pulls strawberries from his basket and places them, one by one, onto the strawberries pictured on the counting strip. Four-year-olds Emma and Lucia stand by a meter stick taped to the wall. They will use this to investigate how stretchy their fabric is. Together they stretch a fabric strip along the meter as far as it will go—which is as long as the meter. "This one is really stretchy," Emma says.

8.3.4.2 Ages 5 and 6

Kindergarten science should similarly bridge disciplines and involve exploration and inquiry (NSTA, 2002). Children are capable of asking questions, conducting investigations, interpreting data, and engaging in other scientific practices at an age-appropriate level, with the support of the teacher and curriculum (Jirout & Zimmerman, 2015). Well-designed engineering challenges can afford excellent opportunities for exploration of science content and engagement in scientific practices in contexts that make sense to children (Brophy et al., 2008; Katehi et al., 2009; Roth, 2001; Zubrowski, 2002). Children can to draw upon knowledge or experiences in science to inform their engineering designs.

> After the children have played and explored outside in the wind, Ms. Martinez gathers them on the rug for a class discussion. "Tell me what you learned about air," she says, red marker poised to write on the easel. "Air can push you," Aliya offers. "Air is something," Malcolm adds, "but you can't see it." Ms. Martinez prompts, "What is another word to describe something you can't see?" "Invisible!" several children call out.

Kindergarteners, like preschoolers, are expected to practice counting skills and to compare things using attributes that are measureable. They work on place value skills, as well as the foundations of adding and subtracting (Common Core State Standards Initiative, 2012b; Department for Education and Skills, 2006; NAEYC

& NCTM, 2010). In Ms. Martinez's room, the most salient use of mathematics is in measurement and comparing, as children measure how far the boat travels by counting tiles on the floor along the sail track.

> Malcolm gives his tin-foil sail, taped to the craft stick mast, to Ms. Martinez at the testing station. She places it in the foam boat and turns on the fan. The boat zooms along its filament track, stopping about half way. Malcolm jumps up and down in excitement. "This one is much better than my last one!" he says. "How much better?" asks his teacher. Malcolm quickly counts tiles on the floor beneath the track. "This one went eight tiles. My last sail only went three tiles!"

8.3.4.3 Ages 7 and 8

After kindergarten, children are expected to grow more skillful with science practices and accumulate more content knowledge. They are expected to become more capable of constructing arguments from evidence, for example (Jirout & Zimmerman, 2015; NGSS Lead States, 2013). Ms. Chen conducts a whole-class discussion, asking children to compare the materials they will test to other pollinators they know.

> "Which materials do you predict will work the best?" Ms. Chen asks. "The pom-pom," answers Maria. "Why?" Ms. Chen prompts. "Because it's fuzzy, so the pollen will get stuck in it." "I think the pipe cleaner will work better," says Ariana, "because it's more like a bee leg. It has little hairs on it that will hold the pollen."

Mathematical expectations for children in the 7-to 8-year-old age band include further work with addition and subtraction using a variety of strategies and models, development of fluency with base-ten operations, investigation of shapes and their composition, further work with measurement using both standard and nonstandard measurement, and the representation and interpretation of data (Common Core State Standards Initiative, 2012b; Department for Education and Skills, 2006). In Ms. Chen's class, children organize data from testing, first in their engineering journals, and then as a class on a summary table. As described above, they discuss the testing results and notice discrepancies, leading them to question and revise their testing methods.

8.3.5 Analysis of Data for Planning and Redesign

Even the youngest preschoolers have the ability to engage in scientific practices (Jones et al., 2008; NSTA, 2014), and the same applies to engineering practices. With the help of adults to scaffold children's practices (Jirout & Zimmerman, 2015; Quintana et al., 2004) and engage them in the zone of proximal development (French & Woodring, 2014), children can develop their skills over time. One such practice is the ability to test, collect data, and evaluate results.

In engineering, data analysis is used to inform the planning process and also in the process of testing and improving candidate solutions. Like engineers, children can

Table 8.6 Developmental trajectory for design parameter 5: analysis of data for planning and redesign

What it looks like: ages 3–4	What it looks like: ages 5–6	What it looks like: ages 7–8
• Children test materials and evaluate them on a two-dimensional scale (e.g., louder/quieter) • With teacher support, children construct simple charts or compare results through whole-group discussion • Children judge the success of their design solution through observation and comparison against a standard, low-quality solution • Children are encouraged to improve their designs repeatedly	• Children test materials and evaluate them for specific qualities • With teacher support, children construct graphs and charts and discuss and compare results across the class to draw lessons for planning a design solution • Children judge the success of a design solution using a specified testing procedure to make qualitative judgments • Children are encouraged to use their results and findings to improve their designs repeatedly	• Children test materials and methods of construction for specific qualities • With teacher support, children construct graphs and charts and discuss and compare results across the class to draw lessons about "fair tests" and planning a design solution • Children judge the success of a design solution using a specified testing procedure to make qualitative judgments and quantitative measures • Children analyze and describe which parts of their technology failed during testing and offer suggestions for modifications they will make in redesign

collect data about the materials they want to use, about different types of methods and design shapes, and about their design solutions. With enough time to test, evaluate, improve, and redesign, across multiple cycles, they can develop solutions to age-appropriate design challenges (Brophy et al., 2008; Katehi et al., 2009). Repeated opportunities to reflect, put learning and ideas into action, test, and reflect again give children a chance to develop deep and lasting understanding (Sawyer, 2006b).

In the EiE curriculum, we provide a sampling of prompts that teachers can use to encourage children to reflect on data and draw conclusions. Written materials prompt children to record their data, help them to structure it, and prompt children to draw conclusions. For younger children, especially pre-readers, much of the work of scaffolding children's reflection on tests and data must be done directly by the teacher, or by simple visual materials, such as a scale where children can set out items in order from lightest to heaviest, or from quietest to loudest (Table 8.6).

8.3.5.1 Ages 3 and 4

Preschool children must be able to judge success or failure through observation, with a single, straightforward criterion. In the case of the basket, children can observe whether the basket holds fruit.

In Ms. Jones's room, Jackson stuffs his completed basket with plastic strawberries. "Oh no!" he cries, as strawberries pop out between his fabric strips. "There's a hole in my basket!"

Once they achieve this criterion, children might naturally gravitate toward more difficult criteria such as trying to hold more fruit or for longer periods of time, or the teacher can encourage them to do so.

8.3.5.2 Ages 5 and 6

Kindergarten children, like preschoolers, must be able to quickly ascertain whether or not their design is functioning properly. However, they are also ready to begin reasoning about the phenomena they observe and citing evidence (Jirout & Zimmerman, 2015; NGSS Lead States, 2013).

Aliya brings her triangular white sail to the testing station. However, when her mast is inserted into the boat in front of the fan, her sail flaps and the boat doesn't move at all. "It's like a flag!" she exclaims. "That's a good observation. What do you think you need to do differently to get it to work?" Ms. Martinez prompts. "I think," Aliya says slowly, looking carefully at her nonfunctional sail, "that it needs to be like this," (she cups her hand) "to catch the wind."

8.3.5.3 Ages 7 and 8

For children in grades 1 and 2, testing can involve multiple aspects and scoring systems that measure success qualitatively. Children can record their observations and results of tests with pictures and short sentences or fragments. They can discuss simple trade-offs and criteria. They are also ready, and often eager, to discuss "fairness" in testing and are ready to be introduced to the concept of controlled tests (Jones et al., 2008; NGSS Lead States, 2013).

Ms. Chen gathers her class to discuss the results of their first round of testing. She asks each group for their results, and writes on chart paper whether they were able to transfer "None, a little, or a lot" of pollen to the paper flower models for each of their trials. One group in particular has excellent results every time. "But that's not fair!" Daniel exclaims, "I saw Ariana tap and tap her pollinator really hard until all the pollen fell off." Ms. Chen nods. "That's a really good point. The way that we test is important. Let's put how we tested on our chart. How did you and Maria do your test, Daniel?" She asks each of the groups in turn to describe how they had tested, and writes it on the chart. "Can we compare these results fairly, class?" "No," the children chorus. "So what should we do? Ariana?" "We should all do it the same way," she offers. "Like, we could all tap like this," she says, demonstrating with her hands. "And we should all tap the same number of times," adds Daniel.

Table 8.7 Developmental trajectory for design parameter 6: collaboration

What it looks like: ages 3–4	What it looks like: ages 5–6	What it looks like: ages 7–8
• Children work alone or in freely formed groups • The teacher leads children to interact during whole-group or small-group discussion • The teacher praises successful interactions and supports children to manage contentious interactions	• Children work in pairs, sharing materials but designing their own solutions • The teacher discusses and models appropriate interactions • The teacher praises successful interactions and supports children to manage contentious interactions	• Children collaborate in pairs or groups of three on a shared design solution • The teacher discusses and models appropriate interactions • The teacher provides support to consider each other's ideas and negotiate shared solutions

8.3.6 Collaboration

Children change dramatically in their ability to work with others between the ages of 3 and 8. Where at age 3, most children engage primarily in associative play—side by side with others, without joint intentions—by age 8, children are capable of collaborating with others, sharing ideas and goals in joint activity (Copple & Bredekamp, 2009). Children's preferences to work with others also grow during this age range. Minority children and girls, especially, are more likely to be engaged in class environments where collaboration is valued and encouraged (Burke, 2007).

Engineering as a discipline involves teamwork and collaboration (Katehi et al., 2009). Any problem of any importance is likely to be addressed by a team of engineers with scientists or others, using diverse viewpoints and strengths to come up with a high-quality solution (Katehi et al., 2009; Sawyer, 2006a). Like engineers, children can produce better solutions if they work together than if they work alone—at least once they have the social skills to collaborate effectively (Solomon & Hall, 1996).

Engineering curricula can support children's social and emotional development at an appropriate level, with careful thought. Teachers need to plan for how to encourage and support children to work effectively together (Wendell et al., 2014). They need to monitor how children are interacting and intervene as necessary to correct antisocial behavior and support children to work effectively together. It also helps to review norms and expectations with the class (Table 8.7).

8.3.6.1 Ages 3 and 4

Preschool-aged children are just beginning to learn to play with others and interact respectfully with others; sharing is an emergent skill, and children's tempers can run high (Copple, Bredekamp, Koralek, & Charner, 2013; National Institutes of Health, 2014). Children need encouragement for effort and feedback promoting their responsibility to develop persistence (Allen & Kelly, 2015). Three-year-olds can manage simple instructions with two or three steps, and 4-year-olds can manage a little more, though they are easily distracted (Centers for Disease Control and Prevention, 2013; Copple & Bredekamp, 2009). Engineering curricula can help children develop their social and emotional skills by providing opportunities for sharing and side-by-side design activities, but teachers will need to remain heavily involved in supporting children's efforts and interactions. They support children's interactions with each other, helping them to behave cooperatively with each other, and teaching them to react respectfully to each other's ideas, desires, and projects (Copple & Bredekamp, 2009).

Jackson and Emma work side-by-side on their baskets. Jackson pulls the box of hair clips to his side of the table, and when Emma reaches to take one he covers them protectively. Ms. Jones, passing by, notices the brewing argument. "What's wrong?" she asks. "I need them," Jackson explains. "I don't think you need them all," she tells him. "You can't fit this many clips on your basket! Can you share with Emma, please?" Jackson moves the box back to the middle.

8.3.6.2 Ages 5 and 6

By kindergarten, most children are skilled at cooperating with others, though they may be overly assertive or have difficulty following others' lead (Copple & Bredekamp, 2009). They are more independent and persistent, yet they may still need encouragement in the face of a challenge (Allen & Kelly, 2015; Copple, Bredekamp, Koralek, & Charner, 2014; Massachusetts Department of Early Education and Care, 2015). Children can work in pairs or small groups sharing materials and ideas, but each child should be allowed to develop his or her own designs independently.

"This tin-foil sail worked really well," Malcolm tells Aliya, "but I think it should be bigger to catch more wind." He pulls the tin foil from their box and rolls the edges of one sheet to make a large, stiff oval. "Could I have some tin foil?" Aliya asks. Malcolm hands back the rest of the stack of sheets. "I want to try tin foil too, but I'm going to make a frame for it," she tells him, as she gets to work.

8.3.6.3 Ages 7 and 8

Seven- and eight-year-olds are beginning to learn to collaborate and work together. Sharing ideas is still a developing skill, but children in grades 1 and 2 are capable of arguing for their own ideas, taking into account the ideas and desires of others, and compromising (Copple & Bredekamp, 2009). In working on an engineering design project, pairs and sometimes trios can work together on the same design. They need teacher support to collaborate jointly—children frequently find it difficult to agree on a plan for a design.

Daniel and Maria compare the ideas they've each recorded in their engineering journals. "So, let's use idea number three, but we'll just bend it, how about that?" Maria proposes. "So, it would be half of my idea," Daniel concurs. "It will look like your idea, but it will do the things that are in my idea." "Mmhmm," Maria agrees, "so it's both of ours mixed up."

Table 8.8 Developmental trajectory for design parameter 7: agency

What it looks like: ages 3–4	What it looks like: ages 5–6	What it looks like: ages 7–8
• The teacher may directly support the process with prompts and structure • Children make their own choices and engage directly with the materials • Simple visual materials support children to focus on important aspects of materials and the problem	• The teacher models for children and prompts them to come up with and try new ideas • Children make their own choices and engage directly with the materials • Simple written or visual materials support children to focus on important aspects of the problem	• The teacher models for children and prompts them to come up with their own questions and ideas, as well as to make observations and draw their own conclusions • Children work together to make decisions and plans as a team and to create, test, and improve their ideas • Written materials support children to reflect and make connections through open-ended prompts for short answers and basic observations

8.3.7 Agency

Children need to have the agency to make choices if they are to develop creative solutions. When children have agency to develop their own designs, they are more likely to become emotionally invested in their work. Emotional engagement is a strong predictor of deep, lasting learning (Blumenfeld, Kempler, & Krajcik, 2006; Engle & Conant, 2002; Immordino-Yang, 2015). Children benefit from developing the confidence that they can solve their own problems (Katehi et al., 2009).

Teachers can foster agency by stepping back and allowing children to come up with their own ideas and solutions. This is not the same as letting children "discover" things for themselves: Scaffolding and guidance are important to learning (Hmelo-Silver et al., 2007). However, teachers can push children to come up with their best answers, curriculum can develop learning activities that help children to confront important ideas, and teachers can guide children to think deeply about those important ideas, while celebrating and fostering children's ownership of their design solutions (Table 8.8).

8.3.7.1 Ages 3 and 4

Anyone who has worked with small children has heard the sentiment, powerfully (if ungrammatically) expressed, of "Me do it!" Children strive for independence at an early age. Teachers support children's development by providing an environment which is emotionally safe for failure and which encourages persistence (Australian Government Department of Education, Employment, and Workplace, 2009). When adults "take over" for children, they deprive children of opportunities to learn and develop confidence.

In the following vignette, Ms. Jones encourages children to share and own their experiences, helping Jackson to verbalize for himself what she knows he accomplished and learned from his initial failure.

> The children are gathered on the rug for circle time at the end of the day. "What did you learn about making a basket, Jackson?" Ms. Jones asks. "I made a basket," Jackson replies. "You did, I saw your basket! Can you tell me what you learned? Did your basket work the first time?" "No," says Jackson, "the strawberries fell out." "Why is that? Do you have an idea?" prompts Ms. Jones. "It was too stretchy. It stretched and made a hole," answers Jackson.

8.3.7.2 Ages 5 and 6

Though children who are 5 and 6 years old are still quite young, they appreciate the opportunity to make their own decisions. With the help of a supportive teacher and a strong curriculum providing "training wheels," they can create unique and functional design solutions.

Throughout our vignettes, we see how Ms. Martinez encourages children to take an authoritative role when discussing their ideas and designs. She respects and validates their ideas and contributions by taking them up and encouraging them and other children in the class to work further on their own ideas and the ideas of others. In Sect. 8.3.3.2, Ms. Martinez pushes children to discuss their observations of the properties of materials. In Sect. 8.3.4.2, she again asks children to contribute their observations, this time about air. She also pushes Malcolm to quantify his statement that his new sail "is much better" than the prior one. And, in Sect. 8.3.5.2, she encourages Aliya to make her own evaluation of her sail, verbalizing why it is not working and what she needs to change to improve it.

8.3.7.3 Ages 7 and 8

Children aged 7 and 8 are more capable than younger children. They are better at verbalizing their ideas and better at managing complexity and balancing variables. Still, adults can diminish their engagement and accomplishments by being too quick to step in, by drawing conclusions for children instead of allowing them to draw their own, and by telling instead of listening.

In the vignette in Sect. 8.3.5.3, we saw Ms. Chen guide children to work out, as a class, some issues they were having with differences in procedures for testing materials. In the vignette in this section, we see her use her presence to help children work out a way to improve their design solution on their own. This is the kind of guidance and scaffolding we refer to as fostering agency: prompting and guiding children's contributions, and providing validation for children's efforts by accepting them.

"The pom-pom picks up pollen the best," Mason asserts. "Yeah," Dina agrees, "but how are we gonna get it into the flower?" She picks up the bent PVC pipe that is their model for a bucket orchid. "If we try to put the pom-pom in there, it's gonna get stuck." Mason picks up a pipe cleaner. "Let's tape it to this," he suggests. They get to work.

Ms. Chen quietly joins them, and watches as Dina and Mason test their hand pollinator design on the model of a bucket orchid. The pipe cleaner bunches up as the pompom gets stuck in the PVC model, and they can't push it fully in. "It's not working," says Mason. "It almost works. What could you do to improve?" asks Ms. Chen. "I think we need a stronger handle," says Dina. "Maybe we can twist it together with another pipe cleaner," suggests Mason, "then it would be stronger but still bendy."

8.3.8 Engineering Design Processes and Epistemic Practices

Engineering is, at its core, a discipline that is both creative and deliberative. For children to become familiar with the nature of engineering, they need to be introduced to its epistemic practices and a systematic process of design as a way to structure their engagement (Cunningham, 2018; Cunningham & Kelly, 2017; Lewis, 2005). We do not advocate for indoctrinating children into one process of engineering (which does not exist); instead, children's engineering should move flexibly from investigation to planning to testing and back again, to satisfy children's developmental need to alternate between hands-on manipulation and reflection (Brophy et al., 2008; Hill & Anning, 2001). Engineering practices should be discussed directly, with the importance of careful attention to a quality process emphasized.

There are many ways to approach engineering design. We advocate no single process, but the use of an Engineering Design Process as an approach to structure engineering education for children (Cunningham & Carlsen, 2014; Cunningham & Hester, 2007; Davis, Cunningham, & Lachapelle, 2017). By emphasizing engineering design as a flexible process, we prepare children for the idea that failure and setbacks are expected on the way to continual improvement and eventual success. Flexibility in the design process also provides for children's developmental need to move back and forth between hands-on manipulation and reflection (Brophy et al., 2008; Hill & Anning, 2001). For all grades, the cyclical engineering process should include an Explore and Investigate stage, a Create and Test stage, and a Reflect and Improve stage.

During the phase of Explore and Investigate, children learn more about the problem context, the goal, and the materials and methods available for solving the problem. They ask questions and make predictions. By age 7 and 8, they are ready to plan and conduct investigations and to learn about controlling variables. Even the youngest children can make observations, but older children can seek patterns in their observations, and eventually organize and tabulate data.

To start the Create and Test stage, older children can brainstorm individually, recording their ideas with pen and paper. Younger children may be able to imagine and speak their ideas, or draw pictures of their ideas, but the youngest may need to go straight into playing with materials and attempting to create. It is crucial that the choice of design challenge, materials, and presentation be age appropriate—not too challenging for small hands and limited fine motor skills, but not so easy as to be immediately accomplished; not too many materials and methods as to be overwhelming, but not so few as to constrain creativity. Criteria for success for the youngest children should be straightforward and immediately evident (whether the basket holds fruit when you carry it across the room); older children can use a small number of qualitative or readily quantifiable measures to judge the degree of success for their designs.

The Reflect and Improve stage is vital to learning. This is the part of the process where children are invited to think about what they have accomplished, what they have come to understand, and what is still confusing to them. With the opportunity to put their new learning into action, they are able to synthesize and apply what they have learned, either in creating a new design, or in coming up with a new set of questions to ask (Brophy, 1983; Katehi et al., 2009; Sawyer, 2006b).

Most engineering processes include a stage of communicating the final design. In the early years, we include this as part of the Reflect and Improve stage. Preschoolers can be asked to communicate their designs in small groups or one-on-one with the teacher. They can draw or take pictures of their designs and record their thoughts either with audio or with the help of a teacher to scribe for them, so they can revisit them at a later time. Such images can be posted on the wall near the Engineering Exploration station, or shared with the group, with each child saying a few words. As children grow older, their presentations and their engineering journals become more sophisticated, incorporating multiple views, labels, and instructions (Hertel, Cunningham, & Kelly, 2017) (Table 8.9).

8.3.8.1 Ages 3 and 4

In preschool, teachers scaffold children's engagement in engineering and science practices, helping them to form questions, to make predictions and observations, and to communicate and reflect on what they have learned (Jirout & Zimmerman, 2015). Even the youngest children can learn from seeing each other's designs, and from reflection, with the support of the teacher to ask questions and help children formulate their ideas. Teachers must know their students' particular strengths, needs, and weaknesses, so they can appropriately intervene when a child needs encouragement to persist, or help with fine motor control, or provide more detailed prompts to complete a thought.

8.3.8.2 Ages 5 and 6

In kindergarten, children have more skills and are more capable than preschoolers, but they still need significant teacher support (Copple et al., 2014). At this age, they gain the ability to distinguish between models and the things or events those models represent (NGSS Lead States, 2013). With guidance, they can plan an investigation

Table 8.9 Developmental trajectory for design parameter 8: engineering design processes and practices

What it looks like: ages 3–4	What it looks like: ages 5–6	What it looks like: ages 7–8
• The EDP has 3 steps • Children engage in problem scoping, creating and testing prototypes, and making improvements • Teachers work directly with children to provide verbal scaffolding and model epistemic practices • Materials provide visual scaffolding without the need for reading • Children communicate what they've done with drawings, photographs, and speech	• The EDP has 3 or 4 steps • Children engage in problem scoping, creating and testing prototypes, making improvements, and communicating designs • Teachers provide verbal scaffolding and model epistemic practices for the class • Materials are mostly visual but begin to incorporate symbols • Children communicate their ideas and their designs with drawings, photographs, speech, and labels	• The EDP has 4 or 5 steps • Children engage in problem scoping, brainstorming, drawing up plans, creating and testing prototypes, evaluating to make improvements, and communicating designs • Teachers model for the class and ask open-ended, generative questions to encourage children to actively engage • Materials scaffold all processes through simple prompts • Children communicate ideas, designs, and conclusions with drawings, basic writing, and class discussion

and work with their peers to conduct it. They can collect and record data, make comparisons, identify patterns, draw inferences, make arguments based on evidence, and apply what they have learned (French & Woodring, 2014; NGSS Lead States, 2013). Children continue to need monitoring and support to cooperate with each other.

> Ms. Martinez's class gathers at the rug, carrying their sails that worked best, after everyone has had a chance to test at least three designs. "Hold up your sails that made it all the way to the end of the track," Ms. Martinez tells her students. Aliya holds up a sail made of plastic, billowing between a top and bottom stay taped to the mast; Malcolm holds up a tin-foil sail stiffened into a concave oval by rolling the edges. "What properties made the best sails?" Ms. Martinez asks. Aliya raises her hand. "Lots of material to catch the wind," she says. "Thumbs up if you agree!" Ms Martinez tells the class, and everyone puts up their thumbs. "A sail that's stiff so the air can push on it," Malcolm contributes, and again his classmates raise their thumbs.

8.3.8.3 Ages 7 and 8

By ages 7 and 8, children become able to identify the elements in a system and how they relate to each other. Their skills with other practices increase, especially their abilities to draw inferences and make arguments based on evidence (Jirout & Zimmerman, 2015). However, they still need significant support, including teacher modeling and scaffolding, to engineer. Teachers use whole-class discussion time to prompt children to ask questions, think about what they need to know, understand what is expected of them, consider the implications of data, and communicate their findings. Written materials contain simple prompts, often with choices spelled out, to minimize children's need to compose and write.

In the following vignette, we see Ms. Chen using questions to scaffold children to analyze how their hand pollinators worked and draw conclusions that they can use in their next design.

> "What have we learned so far, children, about what makes a good pollinator?" Hands shoot up. "Fuzzy things like pom-poms and pipe cleaners are the best to pick up pollen," says Aliya. "But our pom-pom was too big to fit in our flower," says Mason. "We had to cut our pom-pom to get it all the way in." Ms. Chen responds, "That's a really good point, Mason. We had three different types of flowers. Did anyone make a pollinator that worked best for every flower?" "No!" the class choruses. "For the poppy we need something that is really big and fuzzy," says Maria. "But for the bucket orchid we need something small."

8.4 Teacher's Role in Scaffolding

As we have discussed throughout this chapter, children need significant scaffolding to help them develop and improve their skills in science and engineering. They also need support to develop behaviors that will help them achieve success, such as cooperation, persistence, and collaboration. Some of this scaffolding can be provided by the curriculum, directly to older children with some skill in reading, in the form of worksheets and engineering journals. The rest must be provided by teachers, who model good questions, arguments, and use of data; who encourage children to persist and help to manage interpersonal difficulties; and who prompt children with open-ended questions to reflect and think more deeply about what they are experiencing.

It is not always easy for teachers to know what to look for in an engineering challenge, or what to ask. Engineering is as new for most early childhood teachers as it is for children. Both curriculum and professional development can support teachers to improve their engineering practice. Curriculum guides can provide step-by-step instructions and sample questions for novices. An overview of the goal of the lesson is also important, as is an outline of how the activity is intended to improve children's skills and understanding (Cunningham & Hester, 2007). However, professional development with teachers practicing and interacting around the engineering skills and content knowledge they are learning remains a vital way to help teachers to improve their pedagogy (Diefes-Dux, 2014).

In our experience, professional development is valuable to teachers when it switches between the student's point of view and the teacher's—what we call "student hat" and "teacher hat." Teachers can put themselves in the shoes of children as learners while they work on an activity and then reflect upon the activity as a teacher, thinking about the likely pitfalls and trouble spots their students will encounter. They also practice asking questions that get children thinking and doing—questions that require more than a one-word answer and that respect and validate children's contributions.

8.5 Discussion

Throughout this chapter, we have focused on a variety of aspects of curriculum and pedagogy that are particularly relevant to engineering education for children, including the Narrative Context (Parameter 1); Goals, Constraints, and Requirements (Parameter 2), which includes developing age-appropriate design challenges and supporting children to define the scope of the problem; Materials and Methods (Parameter 3); the Application of Science and Mathematics (Parameter 4); Analysis of Data for Planning and Redesign (Parameter 5); Collaboration (Parameter 6); Agency (Parameter 7); and Engineering Design Processes and Epistemic Practices (Parameter 8). By examining what children are developmentally capable of at each age, we sought to provide some early trajectories that educators and curriculum developers might use as they consider how to design early engineering experiences for young learners and support children to effectively engage with engineering challenges.

We have focused on what engineering should look like for young children, given our theoretical foundation that learning is both social and constructivist, and based on our experiences designing and implementing formal and informal engineering curricula for children ages 3–13. We have recommended a set of parameters for designing engineering curricula and activities and have shown how those design parameters play out in important aspects of engineering education.

So what should engineering look like for young children, ages 3 to 8? Here, we use our curriculum design parameters to review and further discuss characteristics of the aspects of engineering learning we have set out in this chapter.

8.5.1 Narrative Context

Our first design parameter has to do with context: that the challenge should be set in the real world, relevant to children, and conveyed with narrative. This bears particularly on three aspects of engineering education for the youngest children: (1) the choice of design challenge, (2) the way it is presented to children (setting the context), and (3) the way children evaluate it.

By setting the challenge in a real-world context, it becomes possible to connect with both children's lived experience and the wider world. Even when context-setting uses fantasy elements, such as talking animals or puppets, the reason for the challenge should make sense to children: a basket that will not function, a means to get messages across a gap, or a way to pollinate flowers. Having such a context allows children to bring what they already know to bear—whether that is experiences with collecting colored eggs in an Easter basket, playing with a kite, or watching insects in a flower garden. It also allows children to see a larger purpose in what they are learning, so they can have a better understanding of the adults (or older children) they are working and learning to become.

A larger context also affords connections to reading and communicating. There is a wealth of books, both fiction and nonfiction, about bringing things to a party, carrying things, finding ways to communicate, sailboats and other things that use the wind, and flowers and insects. Children can make up their own stories and role play, draw pictures and artwork, and, when they are older, give reports and recommendations like engineers.

8.5.2 Goals, Constraints, and Requirements

Creativity is afforded by the choice of the engineering challenge, the choice of constraints and criteria for success, the array of materials and methods available to children, and the implementation of the engineering design process. If the challenge is open-ended and the engineering design process is flexible, children have the opportunity to bring new ideas and processes to bear. If the design challenge is constrained and the design process is rigid, children will be channeled into a more routine activity that resembles a craft project more than art or engineering (Brophy et al., 2008; Hill & Anning, 2001; Webster, Campbell, & Jane, 2006).

8.5.3 Materials and Methods

It is important to carefully consider the array of materials and methods to offer to children. Children of different ages have very different abilities to manipulate materials and tools. They also differ greatly in their ability to focus on a variety of properties: Too many choices can overwhelm children, but too few can bore them and constrain creativity.

The choice of materials and methods must be made in tandem with defining the engineering challenge for children, so as to afford a variety of valid solutions. The choice of materials and methods also bears on affordances for applying science and mathematics. Children can explore the properties of materials through observation and experimentation. Testing materials opens up affordances for mathematics.

8.5.4 Application of Science and Mathematics

Just as engineers use mathematics and science to develop solutions to problems, so can children. However, teachers and curriculum designers need to be purposeful about how they connect to science and mathematics. Choosing a design challenge with rich connections to science, such as hand pollinators, enables teachers and materials to specifically integrate science content, with careful attention to how it applies in context. However, it is important to know the expectations for the children who are the target audience. Preschoolers are learning to count and use nonstandard measurement schemes, so a challenge like carrying strawberries affords a contextually appropriate opportunity to practice counting. Testing materials affords older children the opportunity to construct tables and charts and affords all children the opportunity to learn the scientific vocabulary of properties and of investigation.

8.5.5 Collaboration

Even the youngest children can begin to learn the skills needed for teamwork. Though 3-year-olds are still frequently engaging in parallel or associative play rather than playing together, they are ready to learn to share materials, such as those needed to construct a basket. Four-year-olds begin to share ideas, 5-year-olds to cooperate, and older children to work together on a shared goal. Again, we see that learning is a slow process of mastery, supported by community norms and goals as well as by adult scaffolding.

8.5.6 Agency

Giving agency to children to develop and evaluate their own solutions is most important for (1) the choice of the engineering challenge, (2) exploring materials and methods, and (3) evaluating the success of solutions. One important benefit of choosing a design challenge with many possible solutions that children can explore and evaluate on their own is that it gives children the agency, and the responsibility, to be innovative and find a solution that works best for them: functionally, aesthetically, and within their abilities. Agency and active learning are strongly associated with children's motivation to learn (Trundle & Saçkes, 2015). Curriculum designers must work to ensure that the design challenge and the methods and materials to solve it are both open-ended and within the skills and abilities of the children of the target age. The best way to do this is to test ideas for the challenge extensively: within the design group, with teachers of the target age range, and especially with children of the target age range.

8.5.7 Engineering Design Processes and Epistemic Practices

Actively engaging children in engineering processes and epistemic practices is particularly applicable to (1) setting the context of the engineering challenge, (2) exploring materials and methods to solve the problem, and (3) evaluating the success of designs. Because children construct their own understanding and improve their skills through supported and scaffolded practices, they need opportunities to engage in engineering practices at an age-appropriate level. While the youngest children are still learning to form questions and make inferences about design failures, by the age of 8, children are participating in the design of an investigation to answer their questions and balancing multiple criteria for success.

Because children learn through guided and scaffolded participation in the practices of a discipline, curriculum designers and especially teachers need to understand what a developmentally appropriate level of engagement in engineering looks like for their children, and work to help children accomplish that level of skill. Asking questions, making predictions, conducting investigations, coming up with ideas and trying them out, evaluating success, applying what was learned to making improvements, and communicating ideas are all aspects of engineering are accessible to even the youngest children, at some level.

The key to developing and implementing age-appropriate engineering education is understanding that mastery takes time, over a long-term trajectory, and considering how to best support children at the level of which they are capable. One important way to support children's learning of engineering as a practice is to name the practices that children are engaged in, to make explicit to children that they are doing engineering, and to help children to see what level of competence they are striving for. When a teacher helps a preschooler to reframe her question or explains what he learned, the teacher is helping the child not only to expand their verbal skills, but to understand what is expected for a good engineering question or explanation. Stories about or from engineers also help children to envision what they are striving for.

8.6 Conclusion

Young children between the ages of 3 and 8 are capable of age-appropriate engineering practice. They benefit from well-designed engineering education in many ways. They have opportunities to learn more about the world, including engineering problems and solutions within many fields of engineering. They can learn and practice science and mathematics in context, as applications. Engineering education can support children in developing the dispositions and skills to approach problems and failures with persistence, creativity, and an open mind. It can motivate children to learn, as children envision themselves as engineers and are given the agency to make choices and have the responsibility to find a solution.

Engineering education also has the potential to address many aspects of children's development. They develop socially as they work and play with others, negotiating the challenges of shared materials, ideas, and goals. They develop emotionally as they deal with the frustration of setbacks and failures, and persist to success. They develop their motor skills by working with materials and constructing their own designs and creations. They develop academically as they practice mathematics, science, and engineering skills and learn content in these realms. And, they develop as members of a classroom community—and eventually a world community—who work together to solve problems and support each other.

References

Allen, L., & Kelly, B. B. (Eds.). (2015). *Transforming the workforce for children birth through age 8: A unifying foundation*. Washington DC: National Academies Press. Retrieved from http://www.nap.edu/read/19401/chapter/1.

Atman, C. J., Yasuhara, K., Adams, R. S., Barker, T. J., Turns, J., & Rhone, E. (2008). Breadth in problem scoping: A comparison of freshman and senior engineering students. *International Journal of Engineering Education, 24*(2), 234–235.

Australia. Department of Education, Employment and Workplace Relations, & Council of Australian Governments. (2009). *Belonging, being & becoming: The early years learning framework for Australia*. Canberra: A.C.T: Dept. of Education, Employment and Workplace Relations for the Council of Australian Governments.

Bairaktarova, D., Evangelou, D., Bagiati, A., & Dobbs-Oates, J. (2012). The role of classroom artifacts in developmental engineering. Presented at the Proceedings of the American Society for Engineering Education Annual Conference, San Antonio, TX.

Barron, B. J. S., Schwartz, D. L., Vye, N. J., Moore, A., Petrosino, A., Zech, L., et al. (1998). Doing with understanding: Lessons from research on problem- and project-based learning. *Journal of the Learning Sciences, 7*, 271–311.

Blumenfeld, P. C., Kempler, T. M., & Krajcik, J. S. (2006). Motivation and cognitive engagement in learning environments. In R. K. Sawyer (Ed.), *The Cambridge handbook of the learning sciences* (pp. 475–488). New York, NY: Cambridge University Press.

Bransford, J. D., Barron, B., Pea, R. D., Meltzoff, A., Kuhl, P., Bell, P., … Sabelli, N. H. (2006). Foundations and opportunities for an interdisciplinary science of learning. In R. K. Sawyer (Ed.), *The Cambridge handbook of the learning sciences* (pp. 19–34). New York, NY: Cambridge University Press.

Brophy, J. (1983). Conceptualizing student motivation. *Educational Psychologist, 18*(3), 200–215.

Brophy, S. P., Klein, S., Portsmore, M., & Rogers, C. (2008). Advancing engineering education in P-12 classrooms. *Journal of Engineering Education, 97*(3), 369–387.

Brotman, J. S., & Moore, F. M. (2008). Girls and science: A review of four themes in the science education literature. *Journal of Research in Science Teaching, 45*(9), 971–1002.

Burke, R. J. (2007). Women and minorities in STEM: A primer. In R. J. Burke & M. C. Mattis (Eds.), *Women and minorities in science, technology, engineering and mathematics: Upping the numbers* (pp. 3–27). Northampton, MA: Edward Elgar Publishing.

Buxton, C. A. (2010). Social problem solving through science: An approach to critical, place-based, science teaching and learning. *Equity & Excellence in Education, 43*(1), 120–135.

Centers for Disease Control and Prevention. (2013). *Developmental milestones*. Retrieved from http://www.cdc.gov/ncbddd/actearly/milestones/index.html.

Christine, M. C., & Gregory, J. K., (2017) Epistemic Practices of Engineering for Education. *Science Education, 101*(3), 486–505.

Clark, R., & Andrews, J. (2010). Researching primary engineering education: UK perspectives, an exploratory study. *European Journal of Engineering Education, 35*(5), 585–595.

Collins, A. (2006). Cognitive apprenticeship. In R. K. Sawyer (Ed.), *The Cambridge handbook of the learning sciences*. New York: Cambridge University Press.

Common Core State Standards Initiative. (2012a). *Common core state standards for English language arts & literacy in history/social studies, science, and technical subjects*. USA: Common Core Standards Initiative.

Common Core State Standards Initiative. (2012b). *Common core state standards for mathematics*. USA: Common Core State Standards Initiative.

Copple, C., & Bredekamp, S. (Eds.). (2009). *Developmentally appropriate practice in early childhood programs serving children from birth through age 8*. USA: NAEYC.

Copple, C., Bredekamp, S., Koralek, D. G., & Charner, K. (Eds.). (2013). *Developmentally Appropriate Practice: Focus on Preschoolers*. USA: NAEYC.

Copple, C., Bredekamp, S., Koralek, D. G., & Charner, K. (Eds.). (2014). *Developmentally Appropriate Practice: Focus on Kindergartners*. USA: NAEYC.
Cunningham, C. M. (2018). Engineering in elementary STEM education: Curriculum design, instruction, learning, and assessment. New York, NY: Teacher College Press.
Cunningham, C. M., & Carlsen, W. S. (2014). Precollege engineering education. In N. G. Lederman & S. K. Abell (Eds.), *Handbook of research on science education*. Mahwah, NJ: Lawrence Erlbaum Associates.
Cunningham, C. M., & Hester, K. (2007). Engineering is elementary: An engineering and technology curriculum for children. In *ASEE Annual Conference & Exposition*. Honolulu, HI.
Cunningham, C. M., & Lachapelle, C. P. (2014). Designing engineering experiences to engage all students. In S. Purzer, J. Strobel, & M. E. Cardella (Eds.), *Engineering in pre-college settings: Synthesizing research, policy, and practices* (pp. 117–140). Lafayette, IN: Purdue University Press.
Cunningham, C. M., & Kelly, G. K. (2017). Epistemic practices of engineering in education. *Science Education, 101*, 486–505. https://doi.org/10.1002/sce.21271
Davis, M. E., Cunningham, C. M., & Lachapelle, C. P. (2017). They can't spell "engineering" but they can do it: Designing an engineering curriculum for the preschool classroom. *Zero to Three Journal, 37*(5), 4–12.
Department for Education. (2014, March 31). Statutory framework for the early years foundation stage. Department for Education. Retrieved from https://www.gov.uk/government/uploads/system/uploads/attachment_data/file/335504/EYFS_framework_from_1_September_2014__with_clarification_note.pdf.
Department for Education and Skills. (2006). *Primary framework for literacy and mathematics*. London: Department for Education and Skills. Retrieved from http://www.educationengland.org.uk/documents/pdfs/2006-primary-national-strategy.pdf.
Diaz, D., & King, P. (2007). Adapting a post-secondary STEM instructional model to K-5 mathematics instruction. In *ASEE Annual Conference & Exposition*. Honolulu, HI. https://peer.asee.org/3054.
Diefes-Dux, H. A. (2014). In-service teacher professional development in engineering education: Early years. In S. Purzer, J. Strobel, & M. E. Cardella (Eds.), *Engineering in pre-college settings: Synthesizing research, policy, and practices* (pp. 233–258). Lafayette, IN: Purdue University Press.
Duschl, R. A. (2008). Science education in three-part harmony: Balancing conceptual, epistemic, and social learning goals. *Review of Research in Education, 32*(1), 268–291.
Elias, C. L., & Berk, L. E. (2002). Self-regulation in young children: Is there a role for sociodramatic play? *Early Childhood Research Quarterly, 17*(2), 216–238. https://doi.org/10.1016/S0885-2006(02)00146-1.
Engle, R. A., & Conant, F. R. (2002). Guiding principles for fostering productive disciplinary engagement: Explaining an emergent argument in a community of learners classroom. *Cognition and Instruction, 20,* 399–483.
Epstein, A. S. (2006). *The intentional teacher: Choosing the best strategies for young children's learning*. Washington, DC: National Association for the Education of Young Children.
Farkas, G., & Beron, K. (2004). The detailed age trajectory of oral vocabulary knowledge: Differences by class and race. *Social Science Research, 33*(3), 464–497.
French, L. A., & Woodring, S. D. (2014). Science education in the early years. In B. Spodek & O. N. Saracho (Eds.), *Handbook of research on the education of young children* (pp. 179–196). New York, NY: Routledge.
Greeno, J. G. (2006). Learning in activity. In R. K. Sawyer (Ed.), *The Cambridge handbook of The learning sciences* (pp. 79–96). New York, NY: Cambridge University Press.
Henricks, T. (2008). The nature of play: An overview. *American Journal of Play, 1*(2), 157–180.
Hertel, J. D., Cunningham, C. M., & Kelly, G. K. (2017). The roles of engineering notebooks in shaping elementary engineering student discourse and practice. *International Journal of Science Education, 9*, 1194–1217. https://doi.org/10.1080/09500693.2017.1317864.

Hill, A. M., & Anning, A. (2001). Primary teachers' and students' understanding of school situated design in Canada and England. *Research in Science Education, 31*(1), 117(19).

Hmelo-Silver, C. E., Duncan, R. G., & Chinn, C. A. (2007). Scaffolding and achievement in problem-based and inquiry learning: A response to Kirschner, Sweller, and Clark (2006). *Educational Psychologist, 42*(2), 99–107.

Hutchins, E. L. (1995). *Cognition in the wild*. Cambridge, MA: MIT Press. Retrieved from http://books.google.com/books?id=CGIaNc3F1MgC&dq=%22Hutchins%22+%22Cognition+in+the+Wild%22+.

Immordino-Yang, M. H. (2015). *Emotions, learning, and the brain: Exploring the educational implications of affective neuroscience* (Vol. 1). New York, NY: W. W. Norton & Co.

Jirout, J., & Zimmerman, C. (2015). Development of science process skills in the early childhood years. In K. C. Trundle, & M. Saçkes (Eds.), *Research in early childhood science education* (pp. 143–165). Dordrecht, The Netherlands: Springer. Retrieved from http://www.springer.com/us/book/9789401795043.

Jones, I., Lake, V. E., & Lin, M. (2008). Early childhood science process skills: Social and developmental considerations. In O. N. Saracho & B. Spodek (Eds.), *Contemporary perspectives on science and technology in early childhood education*. Charlotte, NC: Information Age Publishing.

Katehi, L., Pearson, G., & Feder, M. A. (2009). *Engineering in K-12 education: Understanding the status and improving the prospects*. Washington, DC: National Academies Press.

Klassen, S. (2007). The application of historical narrative in science learning: the Atlantic cable story. *Science & Education, 16*(3–5), 335–352.

Kolodner, J. L. (2002). Facilitating the learning of design practices: Lessons learned from an inquiry into science education. *Journal of Industrial Teacher Education, 39*(3).

Kolodner, J. L. (2006). Case-based reasoning. In R. K. Sawyer (Ed.), *The Cambridge handbook of the learning sciences* (pp. 225–242). Cambridge, UK: Cambridge University Press.

Kolodner, J. L., Camp, P. J., Crismond, D., Fasse, B., Gray, J., Holbrook, J., & Ryan, M. (2004). Promoting deep science learning through case-based reasoning: Rituals and practices in Learning by Design™ classrooms. In N. M. Seel & S. Dijkstra (Eds.), *Curriculum, plans, and processes in instructional design: International perspectives* (pp. 89–89). Mahwah, NJ: Lawrence Erlbaum Associates. Retrieved from internal-pdf://Kolodneretal2004promotingdeepscilearning-2158936832/Kolodneretal 2004promotingdeepscilearning.pdf.

Kolodner, J. L., Gray, J., & Fasse, B. B. (2003). Promoting transfer through case-based reasoning: Rituals and practices in learning by design classrooms. *Cognitive Science Quarterly, 3*(2), 119–170.

Krajcik, J. S., & Blumenfeld, P. C. (2006). Project-based learning. In R. K. Sawyer (Ed.), *The Cambridge handbook of the learning sciences* (pp. 317–334). New York, NY: Cambridge University Press.

Lachapelle, C. P., & Cunningham, C. M. (2010). Assessing elementary students' understanding of engineering and technology concepts. In *ASEE Annual Conference and Exposition*. Louisville, KY.

Lachapelle, C. P., & Cunningham, C. M. (2014). Engineering in elementary schools. In S. Purzer, J. Strobel, & M. Cardella (Eds.), *Engineering in pre-college settings: Synthesizing research, policy, and practices* (pp. 61–88). Lafayette, IN: Purdue University Press.

Lachapelle, C. P., Cunningham, C. M., Jocz, J., Kay, A. E., Phadnis, P., Wertheimer, J., et al. (2011). *Engineering is Elementary: An evaluation of years 4 through 6 field testing*. Boston, MA: Museum of Science.

Lave, J., & Wenger, E. (1991). *Situated learning : Legitimate peripheral participation*. Cambridge [England]; New York: Cambridge University Press.

Lewis, T. (2005). Coming to terms with engineering design as content. *Journal of Technology Education, 16*(2), 37–54.

Macalalag, A., Brockway, D., McKay, M., & McGrath, E. (2008). Partnership to improve student achievement in engineering and science education: Lessons learned in year one. In *American Society for Engineering Education Mid-Atlantic Conference*. Hoboken, NJ.

Massachusetts Department of Early Education and Care. (2015, June). Massachusetts standards for preschool and kindergarten: Social and emotional learning, and approaches to play and learning. Retrieved from http://www.mass.gov/edu/birth-grade-12/early-education-and-care/curriculum-and-learning/pre-school-and-kindergarten-standards.html.

Miller, P. H., Blessing, J. S., & Schwartz, S. (2006). Gender differences in high-school students' views about science. *International Journal of Science Education, 28*(4), 363–381.

National Association for the Education of Young Children [NAEYC]. (2005). NAEYC early childhood program standards and accreditation criteria: The mark of quality in early childhood education. NAEYC. Retrieved from http://www.naeyc.org/files/naeyc/Position%20Statement%20EC%20Standards.pdf.

National Association for the Education of Young Children, & National Council of Teachers of Mathematics [NAEYC & NCTM]. (2010). Early childhood mathematics: Promoting good beginnings. NAEYC. Retrieved from http://www.naeyc.org/files/naeyc/file/positions/psmath.pdf.

National Institutes of Health. (2014, May 14). Developmental milestones record - 3 years. Retrieved December 14, 2015, from https://www.nlm.nih.gov/medlineplus/ency/article/002014.htm.

National Research Council [NRC]. (2007). *Taking science to school: learning and teaching science in grades K–8*. Washington, D.C.: National Academies Press.

National Research Council [NRC]. (2011). *Successful K-12 STEM education: Identifying effective approaches in science, technology, engineering, and mathematics*. Washington, DC: National Academies Press. Retrieved from http://www.nap.edu/catalog.php?record_id=13158.

National Research Council [NRC]. (2012). *A framework for K-12 science education: Practices, crosscutting concepts, and core ideas*. Washington, DC: The National Academies Press.

National Science Teachers Association [NSTA]. (2002, July). NSTA position statement: Elementary school science. Retrieved from http://www.nsta.org/about/positions/elementary.aspx.

National Science Teachers Association [NSTA]. (2014, January). NSTA position statement: Early childhood science education. Retrieved from http://www.nsta.org/about/positions/earlychildhood.aspx.

NGSS Lead States. (2013). *Next Generation Science Standards: For states, by states*. Washington, DC: The National Academies Press. Retrieved from http://www.nextgenscience.org/.

Oh, Y., Lachapelle, C. P., Shams, M. F., Hertel, J. D., & Cunningham, C. M. (2016). *Evaluating the efficacy of Engineering is Elementary for student learning of engineering and science concepts*. Washington, D.C.: Presented at the American Educational Research Association Annual Meeting.

Palincsar, A. S. (2005). 12 Social constructivist perspectives on teaching and learning. In H. Daniels (Ed.), *An introduction to Vygotsky* (p. 285). New York, NY: Routledge.

Pearson, G. (2004). Collaboration conundrum. *Journal of Technology Education, 15*(2), 66–76.

Penner, D. E., Giles, N. D., Lehrer, R., & Schauble, L. (1997). Building functional models: Designing an elbow. *Journal of Research in Science Teaching, 34*(2), 125–143.

Piaget, J., Gruber, H. E., & Vonèche, J. J. (1977). *The Essential Piaget*. Basic Books.

Quintana, C., Reiser, B. J., Davis, E. A., Krajcik, J. S., Fretz, E., Duncan, R. G., … Soloway, E. (2004). A scaffolding design framework for software to support science inquiry. *Journal of the Learning Sciences, 13*(3), 337–386.

Rogers, C., & Portsmore, M. (2004). Bringing engineering to elementary school. *Journal of STEM Education, 5*(3–4), 17–28.

Rogoff, B. (2003). *The cultural nature of human development*. Oxford, UK: Oxford University Press.

Roth, W.-M. (2001). Learning science through technological design. *Journal of Research in Science Teaching, 38*(7), 768–790.

Roth, W.-M., & Lee, Y.-J. (2007). "Vygotsky's neglected legacy": Cultural-historical activity theory. *Review of Educational Research, 77*(2), 186–232.

Sadler, P. M., Coyle, H. P., & Schwartz, M. (2000). Engineering competitions in the middle school classroom: Key elements in developing effective design challenges. *Journal of the Learning Sciences, 9*(3), 299–327.

Sawyer, R. K. (2006a). Educating for innovation. *Thinking Skills and Creativity, 1*(1), 41–48.

Sawyer, R. K. (2006b). Introduction: The new science of learning. In R. K. Sawyer (Ed.), *The Cambridge handbook of the learning sciences* (pp. 1–16). Cambridge, UK: Cambridge University Press. Retrieved from internal-pdf://Sawyer 2006 New science of learning-3484481024/Sawyer 2006 New science of learning.pdf.

Silk, E. M., Schunn, C. D., & Cary, M. S. (2009). The impact of an engineering design curriculum on science reasoning in an urban setting. *Journal of Science Education and Technology, 18*(3), 209–223.

Solomon, J., & Hall, S. (1996). An inquiry into progression in primary technology: A role for teaching. *International Journal of Technology and Design Education, 6*(3), 263–282. https://doi.org/10.1007/BF00419883.

Stewart, N. (2011). *How children learn: The characteristics of effective early learning*. England: British Association for Early Childhood Education.

Stinner, A. (1996). Providing a contextual base and a theoretical structure to guide the teaching of science from early years to senior years. *Science & Education, 5*(3), 247–266. https://doi.org/10.1007/bf00414315.

Stouffer, W., Russell, J. S., & Oliva, M. G. (2004). Making the strange familiar: Creativity and the future of engineering education (pp. 20–23). Presented at the Proceedings of the 2004 American Society for Engineering Education Annual Conference & Exposition.

The Commission on Early Childhood Education and Care. (2011). *Council conclusions on early childhood education and care: Providing all our children with the best start for the world of tomorrow*. Brussels: European Union. Retrieved from http://eur-lex.europa.eu/legal-content/EN/TXT/?uri=CELEX:52011DC0066.

Thompson, S., & Lyons, J. (2008). Engineers in the classroom: Their influence on African-American students' perceptions of engineering. *School Science and Mathematics, 108*(5), 197–211.

Trundle, K. C., & Saçkes, M. (Eds.). (2015). *Research in early childhood science education*. Dordrecht, The Netherlands: Springer. Retrieved from http://www.springer.com/us/book/9789401795043.

Vygotsky, L. S. (1978). *Mind in society: The development of higher psychological processes*. Cambridge, MA: Harvard University Press.

Webster, A., Campbell, C., & Jane, B. (2006). Enhancing the creative process for learning in primary technology education. *International Journal of Technology and Design Education, 16*(3), 221–235. https://doi.org/10.1007/s10798-005-5633-0.

Wendell, K., Kendall, A., Portsmore, M., Wright, C. G., Jarvin, L., & Rogers, C. (2014). Embedding elementary school science instruction in engineering design problem solving. In S. Purzer, J. Strobel, & M. E. Cardella (Eds.), *Engineering in pre-college settings: Synthesizing research, policy, and practices* (pp. 143–162). Lafayette, IN: Purdue University Press.

Wendell, K., Wright, C., & Paugh, P. (2015). *Engineering design as disciplinary discourse: An exploration of language demands and resources among urban elementary students*. Chicago, IL: Presented at the NARST.

Wicklein, R. C. (2006). Five good reasons for engineering as THE focus for technology education. International Technology Education Association. Retrieved from http://digitalcommons.usu.edu/cgi/viewcontent.cgi?article=1065&context=ncete_publications.

Wilson, E. O. (2002). The power of story. *American Educator, 26*(1), 8–11.

Zubrowski, B. (2002). Integrating science into design technology projects: Using a standard model in the design process. *Journal of Technology Education, 13*(2), 48–67.

Chapter 9
Engineering in Early Elementary Classrooms Through the Integration of High-Quality Literature, Design, and STEM+C Content

Kristina M. Tank, Tamara J. Moore, Brianna L. Dorie, Elizabeth Gajdzik, M. Terri Sanger, Anastasia M. Rynearson and Emma F. Mann

Abstract The *PictureSTEM* project consists of instructional units for grades K-2 that employ engineering and literacy contexts to integrate science, technology, engineering, mathematics, and computational thinking (STEM+C) content instruction in meaningful and significant ways. The *PictureSTEM* project utilizes picture books and an engineering design challenge to provide students with authentic, contextual activities that engage learners in specific STEM content. Four components differentiate the *PictureSTEM* units from what teachers are currently implementing in their classrooms: (1) engineering design as the interdisciplinary glue, (2) realistic engineering contexts to promote student engagement, (3) high-quality literature to facilitate meaningful connections, and (4) instruction of specific STEM+C content

K. M. Tank (✉)
Iowa State University, Ames, IA, USA
e-mail: kmtank@iastate.edu

T. J. Moore · E. Gajdzik · M. Terri Sanger · E. F. Mann
Purdue University, West Lafayette, IN, USA
e-mail: tamara@purdue.edu

E. Gajdzik
e-mail: egajdzik@purdue.edu

M. Terri Sanger
e-mail: mtsanger@purdue.edu

E. F. Mann
e-mail: emann@purdue.edu

B. L. Dorie
Gonzaga University, Spokane, WA, USA
e-mail: dorie@gonzaga.edu

A. M. Rynearson
Campbell University, Buies Creek, NC, USA
e-mail: amrynearson@campbell.edu

L. English and T. Moore (eds.), *Early Engineering Learning*, Early Mathematics Learning and Development, https://doi.org/10.1007/978-981-10-8621-2_9

within an integrated approach. Examples from research data on the *PictureSTEM* unit, *Designing Paper Baskets,* conducted in kindergarten classrooms, will illustrate how the four foundational components of this integrated STEM curricula play an important role in designing meaningful and contextual learning for younger students.

9.1 Introduction

Over the past decade, there has been an increased emphasis on improving science, technology, engineering, mathematics, and computational thinking (STEM+C) teaching and learning at all levels. From this point forward for ease of reading, our use of STEM will be referring to Science, Technology, Engineering, Mathematics and Computational Thinking. In the USA, engineering practices, concepts, and thinking skills are included in the Next Generation Science Standards (NGSS; NGSS Lead States, 2013) as well as within most academic state standards (Moore, Tank, Glancy, & Kersten, 2015). Across the country, early childhood educators are working to integrate these ideas into their classrooms to either meet the standards or ensure school readiness. While there has been an increase in the number and variety of curricula, programs, and specialized schools that have emerged to meet this need for STEM and engineering integration (National Research Council [NRC], 2014), many of these resources have not been focused on early childhood classrooms and settings.

An additional consideration that is important when thinking about the integration of STEM into early childhood classrooms is that there is a large emphasis placed on learning to read. As a result, teachers dedicate substantial amounts of time and energy into ensuring that their students meet this goal (Taylor, Pearson, Clark, & Walpole, 2000). This often leads to a limited amount of time and resources dedicated to STEM subjects (Banilower et al., 2013; Marx & Harris, 2006). As teachers and schools continue to integrate STEM into their early elementary classrooms, it is important to provide them with curricular materials that are well-suited and intentionally designed to provide early learners with appropriate engineering learning experiences. These curricula also need to recognize and consider the challenges and pressures placed upon early childhood educators, like limited instructional time in STEM and a focus on school readiness and learning to read.

The *PictureSTEM* project was developed to meet this growing need to teach young learners high-quality STEM content as highlighted in academic standards (e.g., NGSS Lead States, 2013 and National Governors Association [NGA] Center for Best Practices, & Council of Chief State School Officers, 2010), as well as to immerse students in authentic engineering experiences. This chapter presents examples from the *PictureSTEM* project to highlight how literature and engineering design can be used as a basis for developing engineering, science, mathematics and computational thinking in early elementary classrooms. Classroom examples from the *PictureSTEM* unit, *Designing Paper Baskets,* will be shared to illustrate how foundational components of this integrated STEM curriculum play an important role in designing meaningful and contextual learning for young students.

9.1.1 Overview of PictureSTEM *Project*

The *PictureSTEM* project leverages the traditional emphasis that is placed on reading in early elementary classrooms by integrating children's literature and reading instruction into an integrated approach to STEM using engineering design as a point of focus. The use of high-quality children's literature in *PictureSTEM* is based on the idea that story and context engage student interest and provide a means to integrate learning across disciplines. The other critical component that is used to facilitate student learning in STEM is engineering design, allowing for a real-world context in which students experience the interdisciplinary nature of learning science and mathematics while learning about engineering design and developing engineering habits of mind. The *PictureSTEM* project presents a transformative model for engaging learners in specific STEM content, while also helping to highlight the natural connections across traditional content boundaries.

The development of the *PictureSTEM* units was guided by the framework for STEM Integration in the classroom (Moore, Guzey, & Brown, 2014; Moore, Stohlmann, et al., 2014). This framework suggests that high-quality STEM integration learning experiences have rich, meaningful, and engaging contexts that allow all students to enter into the problem; they include engineering design with an opportunity to learn from failure and redesign based on what they learned; they teach standards-based mathematics and science content using student-centered pedagogies—in particular, evidence-based reasoning as a means to tie the subjects together; they promote teamwork and communication skills; and engineering is threaded throughout the unit, not just tacked on the end. Continued development of this curriculum followed a design-research framework (Hjalmarson & Lesh, 2008) with the current *PictureSTEM* project consisting of three integrated STEM and literacy units for grades K-2: *Designing Paper Baskets*, *Designing Hamster Habitats*, and *Designing Toy Box Organizers*.

The *PictureSTEM* units have a strong theoretical foundation for STEM and literacy integration, with each unit including five paired reading and STEM lessons. In general, the reading lessons provide the students with context, background knowledge, and/or connection concepts. The STEM lessons focus on science, mathematics, computational thinking, and/or engineering content, but always integrate the engineering design context into the content learning throughout the unit. While the units were designed so that the contexts, practices, and content work together to make a learning experience that is realistic and highly engaging for all young learners, the units are also designed to allow for flexibility in classroom implementation. The six paired lessons can be implemented in variety of ways, such as one lesson per day for 12 days or two lessons per day for 6 days and in the block of time that makes sense for school, teacher, and student.

9.1.2 *PictureSTEM* Engineering Design Process

Engineering plays an important role in facilitating learning throughout the *PictureSTEM* units, and within the curriculum, there is a focus on the design aspect of engineering. The representation of the engineering design process (EDP) used within the *PictureSTEM* units (Fig. 9.1) consists of six main components that are divided into problem scoping and solution generation activities and are connected with communication and teamwork throughout. While the design process below uses keywords that are at a developmental level accessible for younger students, it is important to introduce students to the entire phrase (e.g., "Learn *about the problem,*" "Try *the solution*") in order to capture the action associated with each step.

The problem scoping activities require students to **define** and **learn** about the problem and the background information that would be helpful in solving the problem. In the **define** a problem phase, both the criteria (requirements) and constraints (limits) need to be identified, as well as learning about the client (who hired you to do the work) and end user (who will use the end product or process). This information helps students to frame the problem and work toward identifying and evaluating their design solution as they move through the design process. The task of identifying, defining, and better understanding the nature of the problem space is an extremely important aspect of design, especially for younger students who often rush through this step on their way to building and testing their design solutions (Watkins, Spencer, & Hammer, 2014). Within the *PictureSTEM* units, students are introduced to the engineering problem and context at the start of the introduction lesson and then continue to work as a class to define and develop a better understanding of the problem they are trying to solve throughout the lessons. The other important component within the problem space is the **learn** phase, where students are learning more about the problem as they gather scientific and mathematical knowledge and exploring what has already been done to solve the problem. As part of the learn phase, students are asked to research the problem, participate in science and mathematics activities to gain necessary background knowledge, identify constraints attached to the problem, and determine how they will know if their design is successful. Through this problem scoping, students will have gathered knowledge about the problem and content that will help them to be more intentional with making decisions about their designs.

Solution generation within engineering design is also multifaceted. The first facet of solution generation asks students to develop a **plan** for their design solution which includes brainstorming, proposing multiple potential solutions, and evaluating the pros and cons of competing solutions. In doing so, they must evaluate the different constraints on the design as well as establish the relative importance of trade-offs. Students then use the developed plan to **try** out their design through the creation of a prototype, model, or other product.

After a model or prototype is created, it must be tested and evaluated against the criteria and constraints that were determined during problem scoping. As students **test** their solution and determine if their designs are meeting the stated criteria and constraints, students may collect (and/or be provided with) data through various

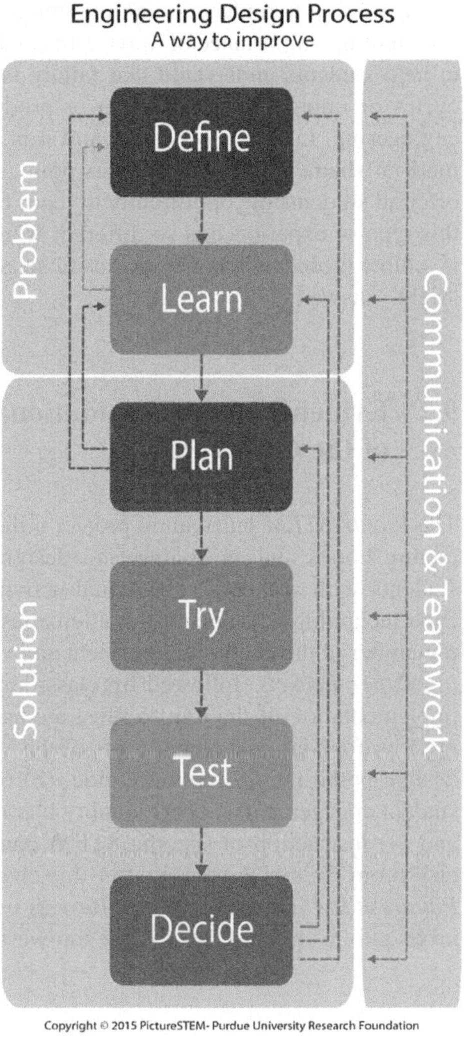

Fig. 9.1 Engineering design process

experiments. These data also help students as they evaluate their prototype or solution based on strengths and weaknesses and **decide** whether their solution is good enough to meet the criteria and stay within the constraints or if they need to use the feedback to redesign their solution. It should be noted that while the engineering design process here is presented to the reader in a linear fashion, the arrows on Fig. 9.1 help the student make decisions on what to do next as they reflect on what they learned from the previous steps taken, not necessarily in a specific order. Helping students to process and reflect on how their design could be improved based on their current findings is an important task as it helps students to better understand the iterative nature of design.

In addition to understanding the importance of iteration, the *PictureSTEM* units highlight the importance of moving through a design process multiple times in order to help students understand that failure is acceptable and to provide feedback for cycles of improving the solution or product until it meets the design criteria. In engineering, failure is expected, and it is used to improve solutions, not seen as a mark of shame. To emphasize this point, all of the *PictureSTEM* curriculum units offer all students the opportunity to learn from failure and participate in redesign as this type of experience is so different from most of what is taught in school—that is, unlike problems in mathematics or science, engineering designs do not have one "right" answer.

9.2 Evidence of the Foundational Components of the Curricula

The *PictureSTEM* curriculum project utilizes a model for STEM learning that uses picture books and an engineering design challenge to provide early elementary students with authentic, contextual activities that engage learners in specific science, technology, engineering mathematics, and computational thinking content. The remainder of the chapter will present an overview of the *PictureSTEM* unit, *Designing Paper Baskets,* followed by classroom examples from kindergarten teachers' implementation of the unit to illustrate and anchor discussions around the following four foundational components of the PictureSTEM curriculum: (1) engineering design as the interdisciplinary glue, (2) realistic engineering contexts to promote student engagement, (3) high-quality literature to facilitate meaningful connections, and (4) instruction of specific STEM content within an integrated approach. The classroom examples presented in this chapter are from the implementation of the *PictureSTEM: Designing Paper Baskets* unit in three kindergarten classrooms at an urban public charter school in the midwestern part of the USA.

9.2.1 PictureSTEM: Designing Paper Baskets Overview

The *Designing Paper Baskets* unit focuses on the development of the mathematics concept of pattern recognition and the science concept of exploring physical properties and materials situated within a hands-on engineering design task. Students are presented with an engineering challenge in which they assist fellow kindergarteners, Max and Lola, in creating a design for a paper basket that can be used to transport wet and dry rocks for other children interested in starting a rock collection. The unit is separated into six distinct lessons that incorporate a coordinating book and a connected STEM lesson (see Table 9.1).

Table 9.1 Design and layout of the *PictureSTEM: Designing Paper Baskets* Unit

Lesson	Intro/1A/B	2A/B	3A/B	4A/B	5A/B	6A/B
Focus	Understanding the problem/understanding paper **EDP**: Define/Learn	Paper and water **EDP**: Learn	Testing paper strength **EDP**: Learn	Pattern recognition and development **EDP**: Learn	Designing Baskets **EDP**: Plan/try	Basket testing and redesign **EDP**: Test/decide/redesign
Picture book(s)	*If you find a rock and be a friend to trees*	*I get wet*		*Pattern fish*	*The most magnificent thing*	*Rocks, jeans, and busy machines*
Literacy focus	Identification of beginning and ending sounds	Blending of three phoneme words	Summarizing informational text	Rhyming words	Making connections to text	Summarizing narrative text
STEM lessons	Identify properties of paper samples and sort using those properties	Learn about properties of wet and dry paper through the water drop test, wax and water test	Test the strength of dry/wet paper with rocks	Identify and create patterns, explore repeating and alternating patterns, identify weaving pattern for basket plan	Choose papers and complete basket plan, work on debugging solution, build a model basket	Test baskets with wet and dry rocks and communicate solution to clients

At the start of the unit, students are introduced to the challenge along with the engineering design process (see Fig. 9.1) through a series of letters from Max and Lola that engage students in some problem scoping in an introductory lesson. In the first letter addressed to the students, Max and his friend Lola explain their passion for collecting rocks and ask for the students' assistance in thinking about ideas for what they might give away at their nature center table to help others with their rock collecting. After having students share their ideas, a second letter from Max and Lola arrives explaining that they liked the ideas and then asks for help with the development of a paper basket that can meet several different criteria. Max encourages the students to test the different types of papers to find which best meet the design goals. Through these letters, students are working as a class to further define the problem by identifying and conceptualizing the criteria and constraints, a set of goals and rules, as well as working toward an understanding of what an engineer is and the work that they accomplish.

In the first lesson (1A), students investigate rocks and their potential organization by different properties through the book *If You Find a Rock* (Christian, 2008). The simple poetry of the book aligns with pictures of children interacting with rocks to highlight different descriptions and purposes that rocks may have. While reading the story the second time through, the teacher pulls out words and asks for the students to identify the beginning and ending sounds. Being able to identify such sounds is an important precursor to phonemic blending, in which students blend different phonemes together to make a word. For the first STEM integration lesson (1B), the class reads the first ten pages of *Be a Friend to Trees* (Lauber, 1994) to learn about how paper is developed out of trees. After making a connection between the idea of rock collecting and the engineering design challenge set forth by Max and Lola, students dive deeper into the science concept of physical properties as they explore the properties of paper samples that will be used during the design of their baskets.

In lesson 2A, students are introduced to properties of water through the nonfiction science text, *I Get Wet* (Cobb, 2002). Through this text, students build their phonemic awareness skills by blending sounds from three phoneme words together, while also learning about science content around water and liquids. The coordinating STEM lesson (2B) allows students to investigate what happens when water drops are placed on different types of papers by conducting a water drop test on regular and wax-coated paper

Continuation of discussion from the book *I Get Wet* (Cobb, 2002) allows students to construct further understanding of water properties through interactively creating summary sentences in the third lesson (3A). Students then test the strength of different papers that will be available during the design challenge under dry and then wet conditions, categorizing each paper based on its strength (3B).

Students are introduced to rhyming and patterns through the book *Pattern Fish* (Harris, 2000) as they connect the weaving of baskets, which is a part of their engineering design challenge, to patterning through the poem. As students work on pattern recognition, this concept from the book is linked back to the engineering design challenge by identifying various patterns seen in woven baskets understand why alternating patterns are important in this situation, and use these patterns to complete weavings in pairs (4B).

In lesson (5A), students practice high-level talk about text as they learn about a girl who overcomes failure to create the most magnificent thing and then make connections between the story and their own lives. In the coordinating STEM lesson (5B), students identify errors in different weaving patterns as they work on their debugging skills before using what they learned about patterns to decide which pattern to use in their own designs. Students also use the properties of paper to make decisions about which papers to use in their basket design. Then they complete their first prototype design.

In the final lesson (6A), students are introduced to two engineering kids, Pedro and Violet, who explore civil and construction engineering in the story *Rocks, Jeans, and Busy Machines: An Engineering Kids Storybook* (Rivera & Rivera, 2010). The students are asked to summarize the narrative in three sentences describing the beginning, middle, and end of the story. Students test their first prototype designs under both dry and wet conditions and share their results with the class, talking about changes for their redesign. Then students redesign and test their new basket prototypes. Finally, students use what they learned during testing and discussion to write letters to Max and Lola in which they make final recommendations for their basket design.

9.2.2 Engineering Design as the Interdisciplinary Glue

The first foundational component of the *PictureSTEM* curricular units is using engineering design as the interdisciplinary glue to help students apply science, mathematics, engineering, and technology learning to an engineering design challenge. The practice of engineering inherently requires the practitioner to call upon other disciplinary knowledge in order to solve engineering problems. Designing using an engineering design process (EDP) is one of the distinctive ways in which engineers work. To help young students develop an understanding of engineering design and the ways in which engineers approach problems and develop solutions, the *PictureSTEM* units provide students with engineering design challenges that are "(1) highly iterative; (2) open to the idea that a problem may have many possible solutions; (3) a meaningful context for learning scientific, mathematical, and technological concepts; and (4) a stimulus to systems thinking, modeling, and analysis" (NRC, 2009, p. 4).

PictureSTEM employs an engineering design process as an anchor for each lesson within a unit, which helps students make sense of and see the natural connections between subjects. As part of this engineering design experience, students have the opportunity to do problem scoping (define and learn about the problem) as well as work toward solution generation (plan, try, test, and decide on a solution). An example of this learning within the *Designing Paper Baskets* unit is seen after the students tested the properties of the paper by looking at how the different papers

Fig. 9.2 Teacher prompt using the engineering design process chart

held up to water. Following the testing, the class discussed which papers they would consider to use in their basket designs, agreeing that tissue paper would not be a good choice due to how poorly it fared during the water testing. The teacher ended the lesson by saying, “Be thinking about that because pretty soon, we’re going to be thinking about designing our baskets, and we’re going to have to use that information that we learned.”

Furthermore, an emphasis is placed on the flexibility of the engineering design process during *PictureSTEM* units. At any point in time, students and the teachers make decisions about what they need to do next based on their current ways of thinking about the problem, the context, and the content. For example, in *Designing Paper Baskets*, a large poster version of EDP with moveable slider is pinned to a wall or an easel for teacher demonstration, and students have small version on their desks with a paperclip to mark their current stage in the process. Students use the EDP to identify where they are in their process to solve the client’s problem. The teacher and students use this tool in multiple ways as they solve their engineering design problem. The teacher may introduce a lesson for the day saying, “Now, we are going to make a Plan,” or “Today, we are in the process of…?” “Learn!” “Learn, that’s right.” as she moves the pointer to the stage of the EDP they will be engaging in (Fig. 9.2). Students may recognize a transition in steps, such as Erica saying, “Can we move our clip to ‘Try’ it?”

Through connection with the EDP, students can recognize how the lessons for science and reading are incorporated into the design challenge. Thus, engineering design provides a foundation upon which students can learn about engineering and problem solving, but it also serves as the "glue" that facilitates learning in other content areas.

9.2.3 Realistic Engineering Contexts to Promote Student Engagement

The second foundational component of the *PictureSTEM* curriculum is the use of realistic contexts to promote student engagement. The context in which engineering design takes place helps to shape the decisions that are made, the criteria for successful completion, and the constraints for the problem that students are solving. This context is made up of the physical and conceptual structure of the problem, the reason for solving the problem, and the social environment in which the problem occurs (Rogoff, 1984). Thus, it is important to use an overarching storyline (i.e., context) to provide a realistic situation for anchoring the work of the young learner. Good contexts for early engineering curricula are engaging (i.e., the context gets them excited to participate in the learning activities), realistic (i.e., the context represents how engineers really work and is meaningful to students), and developmentally appropriate [i.e., the context is understandable to students and within their zone of proximal development (Vygotsky, 1986)]. Each unit within the *PictureSTEM* curriculum includes an intentional and overarching storyline as a context for students that was designed to incorporate the considerations of being engaging, realistic, and developmentally appropriate. An example of how the context and these three considerations play out in the *PictureSTEM* curriculum will be further explained in the following sections using the *Designing Paper Baskets* unit as an example.

9.2.3.1 Engaging Contexts

Engineering contexts engage students through providing a challenge, making personal connections, and providing different entry points to the problem. Young children are already predisposed to solving challenging problems, which in turn can inspire further interest (Carlson & Sullivan, 2004). However, that may not be enough to engage all students. Bringing a more humanistic side to engineering, such as the ability to help others, has been shown to be more personally meaningful for children from historically underrepresented populations in engineering (Hynes & Swenson, 2013). Furthermore, the context of the problem also helps students engage in learning and forges connections to personal interests (Bers, Ponte, Juelich, Viera, & Schenker, 2002) while also providing multiple ways to enter into the problem. In *Designing Paper Baskets*, students have several different hooks that may help them personally engage with the engineering problem they are trying to solve. These different hooks will engage different students. When working together these students will attend different aspects of the problem, which in turn promotes divergent thinking in solution generation. For example, here we see one student engaged with the criterion of the basket looking nice as well as the importance of the strength of the material".

Teacher: *Tell me, what did you choose?*
Sandy: [points to ABBABB pattern]
Teacher: *Why did you choose that?*
Sandy: *Because it looks like a creeper face*
Teacher: *What type of material did you choose?*
Sandy: *I chose wax.*
Teacher: *Why did you choose wax paper?*
Sandy: *Because it's the strongest.*

In the above vignette, the teacher's questions about the student's choice of pattern and materials revealed different aspects of the challenge that she was using as motivation for her design, with pattern choice due to aesthetics and materials due to strength. Using a context that allows students to find their niche in the engineering design project will help them engage with the material. Within the context of *Designing Paper Baskets*, the students might like Max and Lola or potentially relate to them because either they or some of their own friends: (1) like rocks or rock collecting, (2) like weaving or making baskets, (3) like to help people, (4) like to work with their hands, or (5) like to solve problems. In this way, the context provides an additional touchpoint for the students to engage with or reason about the problem.

9.2.3.2 Realistic Contexts

The realistic nature of the context (or overarching story) within the unit is very important for allowing the students to more deeply engage within the engineering design challenge. Engineering challenges that are structured for a design goal provide an end-in-view for students as they work toward solving the problem (English & Lesh, 2003; Lesh, Yoon, & Zawojewski, 2007). Emphasis on the process of engineering design (including testing and iterating), working for a client, and considering criteria and constraints as a way to solve the problem are just several ways to highlight the real work that engineers do and allow for connections to engineering as a profession (Moore et al., 2015). A *Designing Paper Baskets* example that highlights a connection to the work of engineers is seen in the first lesson of the unit. Here, the teacher reminds students that they are solving a specific problem to meet their client's needs with their basket design.

Teacher: *Do you remember the e-mail from Max and Lola?*
Darron: *Yeah.*
Teacher: *What was the problem?*
Darron: *They needed more rocks.*
Teacher: *What?*

Darron: *They needed to build a basket for their rocks.*
[additional class discussion]
Teacher: What was something else about the baskets?
Fred: *We had to figure out what type of strong paper for wet rocks and dry rocks.*
Teacher: *What else about the baskets?*
Fred: *It had to be nice*

In the above vignette, we see the teacher representing the work of engineers through the context of solving a problem and then tying it back to the clients' needs. Within curricula, having clients and end users who "help" students define the problem space provides structure to the problem to allow for greater problem scoping, but also allows for flexibility in terms of thinking about design solutions. The inclusion of a client, in this case Max and Lola, also allows for students to understand the problem from multiple perspectives ranging from the client's needs to how the end user will interact with their final product (e.g., someone who will use the paper basket to collect rocks). The use of context introduces students to and engages them with conceptions about engineers and engineering, such as solving realistic problems, considering criteria, and designing solutions that meet the needs of their clients.

9.2.3.3 Developmentally Appropriate Contexts

Contexts that young learners engage with must also be developmentally appropriate. The importance of developmentally appropriate practices is well-documented within the early childhood literature (Bredekamp & Copple, 2009). Within an engineering design task, the use of a context that is developmentally appropriate is important for facilitating student learning and engagement. For example, young students have developed sophisticated ways of thinking about the world that are based largely on their own experiences with the world (Baillargeon, 2004; Cohen & Chashon, 2006) and providing students with a context that builds upon these experiences is an important step in making the content developmentally appropriate (Bredekamp & Copple, 2009). The use of developmentally appropriate contexts allows for two things within the *PictureSTEM* curricula: helping students navigate the complexity of realistic engineering problems and engaging students in tasks that are at the edge of their zone of proximal development.

Developmentally appropriate engineering provides scaffolding for younger learners. Engineering design problems are generally complex in nature and not well-defined, which can be difficult for young students. These types of problems often have large numbers of criteria and constraints that can make it difficult for students in terms of the bounding of these problems (Jonassen, Strobel, & Lee, 2006). For example, competing variables within the design provide opportunity for students to make decisions, but also the opportunity for students to get confused. Well-designed contexts can help to scaffold these complex and open-ended problems by providing boundaries that allow young children to navigate complexity and see the most relevant issues.

Context also engages students' natural desire to solve problems that are at the edge of their developing capacities (Bredekamp & Copple, 2009). Using a developmentally appropriate context provides a way to scaffold and allow students to engage in tasks that might otherwise be just beyond their reach—for example, designing a device that automatically sorts rocks into different categories is beyond what early elementary students can comprehend but creating a basket to carry rocks is more accessible for students when provided with the necessary scaffolding for the design. Not only does the context provide a way to scaffold for young learners who are new to design, but it allows them to work through problems that are not constrained to a single correct answer (English & Lesh, 2003).

9.2.4 High-Quality Literature to Promote Engagement, Facilitate Integration, and Support Learning

The third foundational component of the *PictureSTEM* curricula is the use of high-quality literature to facilitate meaningful connections. The *PictureSTEM* units recognize the importance of learning to read that is a focus of early elementary classrooms and builds upon this reality by integrating children's literature and reading instruction into an integrated approach to STEM. The use of high-quality fiction and nonfiction literature in *PictureSTEM* is based on the idea that the story and content within the literature can be used to engage students, practice disciplinary strategies, and provide a means to integrate learning across disciplines (Guthrie et al., 2004; McCormick & Hynes, 2012; Palincsar & Magnusson, 2001). The literature component builds upon the idea that the integration of high-quality STEM-focused literature not only supports the learning of literacy skills (Palincsar & Duke, 2004; Yore, 2004), but also supports student learning in other areas by providing background knowledge and a real-world context that is motivating and engaging for students (Cervetti, Pearson, Bravo, & Barber, 2005). Additionally, research in STEM and reading integration has found an increase in student achievement and motivation in multiple areas when integrating reading into science or mathematics instruction (Morrow, Pressley, & Smith, 1997; Palincsar & Magnusson, 2001; Romance & Vitale, 2001; Smolkin, McTigue, Donovan, & Coleman, 2009; Thiessen, 2004), which further supports the integration of picture books in *PictureSTEM* units.

The *PictureSTEM* curricula intentionally tie reading and STEM lessons through a common theme of facilitating learning in STEM without impinging on necessary reading knowledge and skill development. Therefore, the use of high-quality literature in the *PictureSTEM* units was designed to promote engagement, support learning in STEM, and help students to see the connections between the various content areas. The rest of this section will present examples of how high-quality literature is used to promote engagement, facilitate integration, and support learning within the *PictureSTEM: Designing Paper Baskets* unit.

9.2.4.1 Literature to Promote Student Engagement

The use of literature in the *PictureSTEM* units helps to engage students in the unit context and to develop a better understanding of the problem that they will be solving as part of the design challenge. This goal of using the literature to support student engagement is intentionally woven throughout the *PictureSTEM* units. For example, in the *Designing Paper Baskets* unit, the students are introduced to the story of Max and Lola and how they need help in designing a basket that can carry the rocks they find and want to add to their collection. To further promote student engagement, particularly around the idea of collecting things like rocks, the teacher asked students about the different types of things they like to collect before reading the story *If You Find a Rock* (Christian, 2008). The following excerpt provides an example of how the literature is used to engage students in the context of collecting things that is woven throughout the unit.

Teacher:	*Put your hand in the air if you like to collect things.* (students raising hands). *What kind of collection do you have, Sandy?*
Sandy:	*Collection of Nature*
Teacher:	*Oh, you like to collect nature things. What about you* (Alyssa)*?*
Alyssa:	*I have a key collection*
Teacher:	*What about you, Melissa?*
Melissa:	*A rock collection.*
Teacher:	*Susie, what about you?*
Susie:	*I have a flower collection.*
Teacher:	*Do we have any collections in our room?*
Melissa:	*Teacher, it looks like you have a picture collection.*
Teacher (laughing):	*Yes, I have collected a lot of pictures that kids have colored for me. If you look over to our junk boxes, those are all collections. We have collections of frog things, collections of pompoms, buttons and those are all collections.*
Teacher continues:	*Remember, we are solving a problem with Max and Lola and they have a collection of rocks. Today, we are going to read a story called* If You Find a Rock *and we are going to read about it because this is learning about some of the things that we need for our engineering design process*Engineering design process.

Within this example, the discussion around objects they collect precedes their reading of the book about collecting rocks and helps to set up the opportunity for students to connect the point of the picture book (i.e., collecting rocks) to some things that are meaningful to them.

The literature also helps to re-engage students in the unit context prior to participation in the culminating engineering design challenge. Even though the curriculum is designed to have the teacher revisit this context within the STEM lesson, the *PictureSTEM* curriculum also uses the final literacy lesson as a place to remind students of how the context fits into the final design challenge. This is an important step as research suggests that a good understanding of the engineering problem is foundational in helping students to be more intentional with their designing of solutions (NRC, 2012; Watkins, Spencer, & Hammer, 2014). The following excerpt shows the teacher introducing the last book and helping students recall the purpose of the design problem.

Teacher: *We are going to look at this book called* Rocks, Jeans and Busy Machines, *and it is an engineering kids storybook.*
Melissa: *We are going to be engineers?*
Teacher: *We have been being engineers, haven't we? We have been engineering…*
Sandy: *Baskets*
Teacher: *Baskets that can hold wet and dry rocks. We are helping solve a problem for our friends Max and Lola. Well in this story, there is a girl who has adventures as an engineer.*

Here, the literature serves to encourage students' excitement about "being engineers," as well as reminding them of the problem that they will be solving at the end of the unit.

9.2.4.2 Literature to Promote Connections Between STEM and Reading

The literature component in the curriculum is also designed to help students make connections between what they are learning in the reading lessons and in the related STEM lessons. The *PictureSTEM* units, especially the literature and reading lessons, are designed to promote these types of intertextual connections where students are encouraged and given opportunities to reference previous texts and/or experiences while they are currently engaged in a different text (Pappas, Varelas, Barry, & Rife, 2003). This helps students view their learning in a more holistic and less siloed way as they are connecting ideas, learning, and thinking between the reading and STEM lessons, and therefore across traditional content boundaries. In the following vignettes, you can see examples of the teacher helping her students to make connections between the reading lesson and STEM lessons. The first example illustrates how the STEM experience can be connected during a literacy lesson and then second example presents how the literacy ideas can be connected during a STEM lesson.

Example 1 STEM connected during reading:
Prior to this lesson, the teacher stopped at the end of the Day 2 reading lesson to perform the activity suggested in the book with students. This section starts with her picking up the book at the start of the next reading lesson (Day 3) to continue learning about the properties of water. During the reading lesson, the teacher stops reading to refer back to what they had done in the previous STEM lesson. In this section, the words of the teacher are both directly from the book, *I Get Wet*, and her own, as noted.

Teacher (reading from book): *There is another reason why water can wet you, can you guess what it is? Let's do another experiment to find out. Get a piece of wax paper and put it under a faucet, take it out from that faucet and touch where the water was, is it wet?*
Several students: *Noo!*
Teacher: *Surprise, the wax paper is dry.*
Teacher (pausing from reading to address the class): *We didn't put it under a faucet but we put water onto it with a dropper, right?*
Student: *Yep!*
Teacher (goes back to reading): *Put a large drop of water on the wax paper, lift the paper up at one end. The drop slides around, can you get it to slide off the wax paper without wetting it?*
Teacher (addressing class again): *Were you guys able to do that without wetting it?*
Student: *Yeah!*
Teacher: *You bet.*
Teacher (continues reading): *Water doesn't...*

Example 2 Reading during STEM:
Teacher: *Alright, earlier today we read* Pattern Fish *and learned about patterns. Why did we learn about patterns?*
Harper: *There's patterns on our basket.*
Teacher: *Because our basket is going to have patterns on it, that's right. We have learned about paper and what happens when paper gets wet and different kinds of paper and how strong they are and tested them and we learned about patterns. Now we are ready to plan. We get to plan a design for our basket. I want you to see.. I have a weaving here that is made of two different papers here, it's got the base paper and another paper.*
Susie: *It looks like checkers!*
Teacher: *It does kind of look like a checkerboard pattern. What patterns do you see?*

In both of these examples, the teacher is using the literature to promote links between the STEM and reading lessons to help students to explicitly see that there are connections between what they are reading and the activities they are doing during their STEM lessons.

9.2.4.3 Literature to Support Learning in STEM

In addition to using the literature to introduce and engage students with the context of the unit and help them make connections across the disciplines, the *PictureSTEM* curricula intentionally promotes the use of literature to support STEM learning and learning objectives throughout the units. The selection and use of high-quality literature in this curriculum enhance and extend content area learning by building background knowledge, infusing more explanatory thinking, and reinforcing concepts that are addressed during the related STEM investigations. This can lead students to experience greater overall growth and understanding of science and mathematics concepts (Cervetti & Barber, 2009; Palincsar & Duke, 2004; Ford, 2004).

In the following example, the teacher starts the lesson by reviewing the different water drop tests that they did in their previous STEM lesson and then moves into the literacy lesson that is supporting their learning of the science concepts of properties and water.

Teacher (holding the book): *We started this book,* I Get Wet, *and remember we are learning. We are in the learning part of the engineering design process and we are learning about water and different types of paper and how water affects paper and that is important because we want to make a basket…*
Susie: *That can hold wet and dry rocks!*
Teacher: *Wet and dry rocks, so if there's wet rocks then we need to know what happens if the paper gets wet. We started reading* I Get Wet, *by Vikki Cobb.*
Teacher (starts reading): *Know the fastest way to cool off on a hot summer day?*
Alyssa: *Get wet!*
Teacher: *You get wet! Know the easiest way to get clean?*
Susie & Alyssa: *You get wet!*
Teacher: *You get wet! Know what happens when you stay out in the rain? You get wet! Water is the stuff that wets you. It is quite amazing. You can see it. You can feel it. But can you answer this question? What shape is water?* (turns the page)
Susie: *I don't know?*
Teacher: *Here's how you can get your answer. Pour it into a glass. What shape is the water? Pour it from the glass into a bowl. Now what shape is the water?* (she continues reading).

Similar examples of introducing or reinforcing other STEM concepts and learning that is tied to the mathematics content (Lesson 4A) and engineering (Lesson 5A) can also be seen in the literacy lessons that precede the STEM lessons focusing on mathematics and engineering. Therefore, the literature within the *PictureSTEM* curriculum provides a way to make connections but also supports STEM learning across individual units.

9.2.4.4 Literature to Support Learning in Reading

One of the final goals of the literature component in the *PictureSTEM* curriculum is the use of high-quality fiction and nonfiction literature to promote the learning of reading skills and strategies as well as subject matter knowledge. While the overall intention of the curriculum is to use engineering and literacy contexts to integrate STEM content instruction, it is important to note that in order to be meaningful and authentic with the integration of STEM and literacy, the *PictureSTEM* curriculum was also designed to meaningfully teach reading strategies during the literacy lessons. The use of the high-quality literature in this curriculum plays a big role in delivery and facilitation of the reading instruction as each lesson is designed to target a developmentally appropriate comprehension strategy, teach vocabulary at the point of contact, and encourage the use of higher-level thinking, all of which have been shown to help student comprehension in reading (National Reading Panel [NRP], 2000). In this excerpt, the teacher helps the students write a summary about the story and works with students to sound out the letters while constructing one of their sentences.

Teacher: *Afterward, we are going to write some of the beginning, middle, and ending actions in this story.* (the teacher reads the story)

Teacher: *Let's think for a minute about the beginning, middle, and end of the story. So, what happened first in the story, what is the beginning of the story?*

Zach: *She woke up and stretched.*

Teacher: *Yes, that is the very first thing that happened. She woke up and started the day. So, first was…. What did Violet do?*

Alyssa: *Wake up.*

Teacher: *Yes, we are trying to make a sentence about what happened. I will write her name, Violet.* (Teacher writes "Violet" on the chart paper at front of the room).

Zach: *woke up*

Teacher: *ok, so what does woke start with, what sound? /w/, /w/oke?*

Students: */w/oke, /w/oke, /w/*

Sandy: *w*

Teacher: *and then…? w/ō/ke, /ō/*

Alyssa: *o*

Teacher: *and the next sound, wo/k/e?*

Zach: *kay*

Teacher: *Because it is a long sound with the o, to say o (/ō/) instead of o (/ŏ/), we will put an e at the end.* (She writes the rest of the sentence, "Violet woke up," on the chart paper). *Ok, what is the next part that happened in the story?*

The example above shows how the teacher incorporated a reading comprehension strategy into a text that she was using primarily as an introduction to engineering before students started their engineering design challenge. Within *PictureSTEM*, the literature component is not designed to replace all reading instruction, but instead to act as a flexible supplement allowing students to use and build their reading comprehension skills within a content area that connects to their STEM integration activity.

In summary, the use of literature in the *PictureSTEM* curriculum plays an important role in promoting student engagement, fostering connections across the STEM and literacy lessons, and supporting learning in reading and STEM content. Thus, the content and contexts from this high-quality literature serve as one of the foundations within the *PictureSTEM* curricula for meaningfully integrating across all four STEM disciplines.

Table 9.2 STEM and reading content area learning that is utilized in each of the lessons within the *PictureSTEM: Designing Paper Baskets* unit

Lesson	Reading	Science	Technology	Engineering	Mathematics	Computational Thinking
1A	**X**	X		X		
1B	X	**X**	X	**X**		
2A	**X**	X		X		
2B		**X**		**X**		
3A	**X**			X		
3B		**X**	X	**X**	X	
4A	**X**			X	**X**	X
4B		X	X	X	**X**	**X**
5A	**X**		X	X		X
5B		**X**	X	**X**	**X**	X
6A	**X**			X		
6B		**X**	**X**	**X**	**X**	X

A black "X" represents a focal area for the lesson, and a gray "X" represents a supporting area

9.2.5 Instruction of Specific STEM Content Within an Integrated Approach

The fourth foundational component of the *PictureSTEM* curricula is the inclusion of appropriate, standards-based mathematics, and science content learning that is situated within the larger integrated approach to the units. Prior to engaging in the solution generation part of the engineering design challenge, students participate in content-specific learning in the science, and mathematics content that will help them in solving the design challenge. *PictureSTEM* units are designed to require students to apply their mathematics and science content knowledge to their design. This integrated approach to mathematics, science, and engineering deepens students' conceptual understanding in science and/or mathematics (Crismond, 2001; Kolodner et al., 2003; Moore, Stohlmann, et al., 2014; NRC, 2009). Table 9.2 provides an overview of the content area learning within STEM and reading that is included in each of the lessons within the kindergarten *PictureSTEM* unit, *Designing Paper Baskets*.

The breakdown of lessons into content-specific learning highlights the fact that, when implementing an integrated approach, students should have multiple opportunities to engage with and learn about specific content from each of the disciplines that are being integrated. In the case of *PictureSTEM*, the lesson objectives include content learning within science, mathematics, engineering, technology, and reading. To reinforce that the specific content learning is meaningful, each of the *PictureSTEM* curricula highlights specific mathematics and science content objectives (Fortus, Dershimer, Krajcik, Marx, & Mamlok-Naaman, 2004; Mehalik, Doppelt, & Schunn, 2008).

An example of how specific STEM content instruction can be embedded within an integrated approach is seen within the *PictureSTEM* curricula. The engineering context is intentionally presented at the beginning; however, students do not move into solution generation until after participation in problem scoping which includes content-specific mathematics and science lessons.

Science content instruction in the *Designing Paper Baskets* unit allows students to learn about physical properties as they think about which paper would be best for use in their baskets. This helps students to have a better understanding of properties of materials as they first consider the physical properties, then build on that idea in the third lesson when they explore how the strength of the paper changes depending on if they are carrying wet or dry rocks. Science content instruction includes having students conduct experiments with different papers (both wet and dry) to test for changes of texture, transparency, strength, and performance. The following vignette illustrates specific science learning that occurs in the *Designing Paper Baskets* unit, where students are exploring the properties of copy paper as compared to wax paper and what happens when they place drops of water on these different types of paper.

Teacher:	*We are going to see what happens when we put a big drop of water on wax paper and then see if we can pour it back into the cup and then see what happens to the wax paper when we put water on it. Do you guys remember how to use a dropper?*
Students:	*Yes!*
Teacher:	(holding up a dropper). *What is the first thing we are going to do?*
Alyssa:	*Suck up it.*
Sandy:	*Squeeze it.*
Teacher:	*Yes, we are first going to squeeze it.*
Several students:	*Squeeze it.*
Zeke:	*Then put it into the water.*
Susie:	*Then let go.*
Teacher:	*That is the tricky part, remembering to let go. I am going to pass one piece of wax paper and a dropper to each group, and you will take turns with one person being the paper holder and the other person will do the dropper first.*
	[She passes out paper and a dropper to groups. Students place drops of water on the paper, moving them around and then folding the wax paper to pour it back into the cup.]

Teacher: *Look at your wax paper. It had water on it, right?*
Alyssa: *Yes, it's dry!*
Teacher: *Were you able to roll the water off the paper?*
Susie: *Yes, it is slippery!*
Teacher: *Do you think the same thing would happen if we used copy paper?*
Zeke: *No, it would be more like, like stuck.*
Sandy: *Let's try it!*
Teacher: *Yes, let's try it. I am going to dry off your tables first before we continue our experiment with copy paper, so that it doesn't get wet. Now we are going to test copy paper to see if you can roll the drop off copy paper and then see what happens.*

The student-centered science learning demonstrated here both serve the purpose of standards-based science instruction appropriate for kindergarten and helps the students work toward their engineering understanding as they learn more and gain background knowledge about the problem.

The students also conducted tests with the different types of paper to learn how the papers react under loading when wet and dry. Strength tests were performed where students counted the number of rocks that a certain type of paper would hold before it broke (up to 20 rocks) when dry and again when wet. After testing all of the papers when dry, water is dropped onto the paper to test how many rocks the paper can hold when it is wet. The class then notes whether the paper has no physical change due to the stress from the rocks, some change, or has failed. The class also judges the paper as strong, medium, or weak based on how many rocks it held and how the paper looked afterward.

Teacher: *We've got four dropperfuls* [of water on the paper], *now we're ready for the rocks!*
Sandy: *Yeah, it's gonna break.*
Teacher: *You think that's what's going to happen? Let's test it!*
Class counts: *One…Two…Three…Four…Five, it broke!*
Student: *It's weak!*

Student: *That was awesome!* Some of the papers, including copy paper, tissue paper, and paper towels, are likely to break under the load of the rocks when they are wet. Other papers, including the wax and construction paper, are unlikely to break when wet. This science learning around properties of paper is later used when the students choose the types of paper they want to use to make their paper baskets. The students are asked to justify their paper choices based on the evidence they collected from learning about the properties of paper. As an example of how students use this content, the teacher asked each pair of students questions about their choices when testing their prototypes.

Teacher: *What material did you guys choose?*
Andrew: *We both picked wax paper.*
Teacher: *Why did you choose wax paper?*

Jason: *It holds…*
Andrew: *Because it's the strongest paper.*
Jason: *Yeah.*
Teacher: *It's the strongest? How is it the strongest?*
[Jason and Andrew talking at the same time.]
Jason: *Because it didn't break.*
Andrew: *Because…because… because it didn't break.*

The intentional integration of science and engineering in this instance demonstrates the connection these students made between the science lesson and the engineering design challenge.

In addition to science content learning, mathematical content learning is an intentional lesson objective within the *Designing Paper Baskets* unit. In this particular unit, students are introduced to patterns, practice identifying patterns, and learn about the importance of alternating patterns before their group decides on which patterns would be best for their own basket designs. In the following vignette, the teacher is having students identify patterns and assign letters to the patterns they have just read about in the book, *Pattern Fish*.

Teacher: *Ok, Now we are going to look at patterns* (flipping back through the book) *and you guys are very good with your patterns. If you were to use letters to represent this pattern – yellow black, yellow black - what would you say?*
Zeke: *ABAB*
Teacher: *Absolutely, ABAB* (pointing to the book and moving her finger along with the pattern). *What about this one? Use the letters to tell me what pattern this one would be. Stripe dot dot, stripe dot dot.*
Carl: *ABB, ABB*
Teacher: *Yes, great. Allison, this one is for you. Chomp chomp munch munch. Chomp chomp munch munch. What letters would you do?*
Allison: *ABAB.*
Teacher: Ok, listen again. Chomp chomp munch munch. Chomp Chomp, AA, Munch Munch, BB.
Allison: *AABB*
Teacher: *Yes, Chomp chomp munch munch.*
Allison: *AABB*
Teacher: *Lily, this one is for you. Bubble bubble pop, Bubble bubble pop.*
Lily: *AAB* (pointing to the pattern in the book as she says the answer)
Teacher: *Fabulous, you have to keep repeating it because when you say it just once it isn't a total pattern yet, you need to say it a couple of times.*
Lily: *AAB, AAB.*

Within this example, you can see that the teacher is not only helping students to identify patterns, but she is helping to deepen their understanding of patterns by helping them abstract to using letters to represent the patterns they can see in the book. After learning about patterns, students are asked to integrate these mathematical concepts into engineering design as they make decisions about the weaving patterns they want to use in their basket designs. In the following example, as students are working on making decisions about two different parts of their design, choosing a pattern and choosing the type of paper, the teacher is asking students about what patterns they chose for their design.

Teacher: *What kind of pattern did you choose?*
Wendy: (pointing to the squares on her basket) *Blue - orange, blue-orange.*

The intentional integration of mathematics and engineering shows how the *PictureSTEM* curricula are designed to teach students discipline-specific concepts, but also facilitate connections between multiple disciplines. Emphasis on specific content learning takes place as part of the problem scoping to reinforce the need for mathematics and science understanding when they participate in solution generation. This helps students to be more intentional with their design solutions and to deepen content learning as they are asked to use and apply this new mathematics and science learning in their engineering solutions.

9.3 Conclusion

The *PictureSTEM* units present a model of STEM learning with a focus on activities and ideas that are interdisciplinary, integrated across STEM content areas and beyond. This integrated approach allows students more time to receive hands-on interdisciplinary instruction in both STEM content and literacy as well as extend student learning by providing background knowledge and a real-world context that is motivating and engaging for students. Research on the implementation of the *PictureSTEM* modules in early elementary classrooms has found that through this integrated approach students are able to gain a deeper understanding of the science, mathematics, engineering, and computational thinking content as well as make connections across the traditional content boundaries and in contexts outside of the original learning (Tank, Moore, & Pettis, 2013). Additionally, the use of literature has been found to provide teachers with a realistic and engaging context in which to situate student learning of STEM concepts.

This chapter presents examples from the *PictureSTEM* unit *Designing Paper Baskets* to illustrate a model for conceptualizing how engineering and literacy can be used as contexts to promote STEM learning in early elementary classrooms. There are four components that form the foundation for the *PictureSTEM* curriculum: engineering design as the interdisciplinary glue, realistic engineering contexts to promote student engagement, high-quality literature to facilitate meaningful connections, and instruction of specific STEM+C content within an integrated approach. These pieces are crucial in helping students to experience a more interdisciplinary approach to STEM learning that reflects the natural interconnectedness of the four STEM disciplines and helps student to develop critical thinking and problem-solving skills that are needed to solve real-world problems.

Acknowledgements We would like to thank the teachers, students, and researchers that helped make this chapter possible. This material is based upon work supported by the National Science Foundation under Grant No. EEC—1442416, IIP—1519387, and DRL—1543175. Any opinions, findings, and conclusions or recommendations are those of the authors and do not necessarily reflect the views of the National Science Foundation.

References

Baillargeon, R. (2004). Infants' physical world. *Current Directions in Psychological Science, 13*(3), 89–94. https://doi.org/10.1111/j.0963-7214.2004.00281.x.

Banilower, E. R., Smith, P. S., Weiss, I. R., Malzahn, K. A., Campbell, K. M., & Weis, A. M. (2013). *Report of the 2012 national survey of science and mathematics education.* Chapel Hill, NC: Horizon Research, Inc. Available via Horizon Research. http://www.horizon-research.com/2012nssme/research-products/reports/technical-report/.

Bers, M. U., Ponte, I., Juelich, K., & Schenker, J. (2002). Teachers as designers : Integrating robotics in early childhood education. *Information Technology in Childhood Education Annual, 2002*, 123–145. Retrieved from http://www.editlib.org/p/8850/.

Bredekamp, S., & Copple, C. (2009). *Developmentally appropriate practice in early childhood programs: Serving children from birth to eight (National Association for the Education of Young Children Position Statement).* Washington, DC: NAYEC.

Carlson, L., & Sullivan, J. (2004). Exploiting design to inspire interest in engineering across the K-16 engineering curriculum. *International Journal of Engineering Education, 20*(30), 372–380.

Cervetti, G., & Barber, J. (2009). Bringing back books. *Science and Children, 47*(3), 36–39.

Cervetti, G. N., Pearson, P. D., Bravo, M. A., & Barber, J. (2005). R*eading and writing in the service of inquiry-based science.* Regents of the University of California. Retrieved from http://scienceandliteracy.org/research/our_approach

Christian, P. (2008). *If you find a rock.* San Diego, CA: Houghton Mifflin Harcourt.

Cobb, V. (2002). *I get wet.* New York, NY: HarperCollins.

Cohen, L. B., & Chashon, C. H. (2006). Infant cognition. In W. Damon & R.M. Lerner (Eds.), Handbook of child psychology, Set, 6th ed. (Chapter 5, vol. 2). Hoboken, NJ: Wiley.

Crismond, D. (2001). Learning and using science ideas when doing investigate and redesign tasks: A student of naive, novice and expert designers doing constrained and scaffolded design work. *Journal of Research in Science Teaching, 38*(7), 791–820.

English, L. D., & Lesh, R. A. (2003). Ends-in-view problems. In R. A. Lesh & H. Doerr (Eds.), *Beyond constructivism: A models and modelling perspective on mathematics problem solving, learning, and teaching* (pp. 297–316). Mahwah, NJ: Erlbaum.

Ford, D. J. (2004). Highly recommended trade books: Can they be used in inquiry science? In E. W. Saul (Ed.), *Crossing borders in literacy and science instruction: Perspectives on theory and practice* (pp. 277–290). Newark, DE: International Reading Association.

Fortus, D., Dershimer, C., Krajcik, J., Marx, R., & Mamlok-Naaman, R. (2004). Design-based science and student learning. *Journal of Research in Science Teaching, 41*(10), 1081–1110. https://doi.org/10.1002/tea.20040.

Guthrie, J. T., Wigfield, A., Barbosa, P., Perencevich, K. C., Taboada, A., David, M. H., et al. (2004). Increasing reading comprehension and engagement through Concept-oriented Reading Instruction. *Journal of Educational Psychology, 96*(3), 403–423.

Harris, T. (2000). *Pattern fish.* Brookfield, CT: Millbrook Press.

Hjalmarson, M., & Lesh, R. (2008). Engineering and design research: Intersections for education research and design. In A. E. Kelly, R. A. Lesh, & J. Y. Baek (Eds.), *Handbook of design research methods in education: Innovations in science, technology, engineering, and mathematics learning and teaching* (pp. 96–110). London, England: Routledge.

Hynes, M., & Swenson, J. (2013). The humanistic side of engineering: Considering social science and humanities dimensions of engineering in education and research. *Journal of Pre-College Engineering Education Research, 3*(2), 31–42. https://doi.org/10.7771/2157-9288.1070.

Jonassen, D., Strobel, J., & Lee, C. B. (2006). Everyday problem solving in engineering: Lessons for engineering educators. *Journal of Engineering Education, 95*(2), 139–151. https://doi.org/10.1002/j.2168-9830.2006.tb00885.x.

Kolodner, J. L., Camp, P. J., Crismond, D., Fasse, B. B., Gray, J. T., Holbrook, J., et al. (2003). Problem-based learning meets case-based reasoning in the middle-school science classroom: Putting a learning-by-design curriculum into practice. *Journal of the Learning Sciences, 12*(4), 495–548. https://doi.org/10.1207/S15327809JLS1204_2.

Lauber, P. (1994). *Be a friend to trees*. New York, NY: HarperCollins.

Lesh, R., Yoon, C., & Zawojewski, J. (2007). John Dewey revisited—Making mathematics practical versus making practice mathematical. In R. Lesh, E. Hamilton, & J. Kaput (Eds.), *Foundations for the future in mathematics education* (pp. 315–348). Mahwah, NJ: Lawrence Erlbaum Associates.

Marx, R. W., & Harris, C. J. (2006). No child left behind and science education: Opportunities, challenges, and risks. *The Elementary School Journal, 106*(5), 467–477. https://doi.org/10.1086/505441.

McCormick, M., & Hynes, M. M. (2012). *Engineering in a fictional world: Early findings from integrating engineering and literacy*. Paper presented at 2012 ASEE Annual Conference, San Antonio, Texas. https://peer.asee.org/21307.

Mehalik, M. M., Doppelt, Y., & Schunn, C. D. (2008). Middle-school science through design-based learning versus scripted inquiry: Better overall science concept learning and equity gap reduction. *Journal of Engineering Education, 97*(1), 71–85. https://doi.org/10.1002/j.2168-9830.2008.tb00955.x.

Moore, T. J., Guzey, S. S., & Brown, A. (2014a). Greenhouse design to increase habitable land: An engineering unit. *Science Scope, 37*(7), 51–57. https://doi.org/10.2505/4/ss14_037_07_51.

Moore, T. J., Stohlmann, M. S., Wang, H.-H., Tank, K. M., Glancy, A. W., & Roehrig, G. H. (2014b). Implementation and integration of engineering in K-12 STEM education. In Ş. Purzer, J. Strobel, & M. Cardella (Eds.), *Engineering in precollege settings: Research into practice* (pp. 35–60). West Lafayette, IN: Purdue Press.

Moore, T. J., Tank, K. M., Glancy, A. W., & Kersten, J. A. (2015). NGSS and the landscape of engineering in K-12 state science standards. *Journal of Research in Science Teaching, 52*(3), 296–318. https://doi.org/10.1002/tea.21199.

Morrow, L. M., Pressley, M., & Smith, J. K. (1997). The effect of a literature-based program integrated into literacy and science instruction with children from diverse backgrounds. *Reading Research Quarterly, 32*, 54–76. https://doi.org/10.1598/rrq.32.1.4.

National Governors Association Center for Best Practices, & Council of Chief State School Officers. (2010). *Common core state standards for English language arts and mathematics*. Washington, DC: Author.

National Reading Panel. (2000). *Report of the national reading panel*. Washington, DC: National Institute for Child Health and Human Development.

National Research Council. (2009). *Engineering in K-12 education: Understanding the status and improving the prospects*. Washington, DC: The National Academies.

National Research Council. (2012). *A framework for K-12 science education: Practices, crosscutting concepts, and core ideas*. Washington, DC: The National Academies Press.

National Research Council. (2014). *STEM integration in K-12 education: Status, prospects, and an agenda for research*. Washington, DC: National Academies Press.

NGSS Lead States. (2013). *Next generation science standards: For states, by states*. Washington, DC: The National Academies Press.

Palincsar, A. S., & Duke, N. K. (2004). The role of text and text-reader interactions in young children's reading development and achievement. *The Elementary School Journal, 105*(2), 183–197. https://doi.org/10.1086/428864.

Palincsar, A. S., & Magnusson, S. J. (2001). The interplay of first-hand and text-based investigations to model and support the development of scientific knowledge and reasoning. In S. Carver & D. Klahr (Eds.), *Cognition and instruction: Twenty-five years of progress* (pp. 151–194). Mahwah, NJ: Lawrence Erlbaum.

Pappas, C. C., Varelas, M., Barry, A., & Rife, A. (2003). Dialogic inquiry around informational texts: The role of intertextuality in constructing scientific understandings in urban primary classrooms. *Linguistics and Education, 13*(4), 435–482. https://doi.org/10.1016/s0898-5898(03)00004-4.

Rivera, A., & Rivera, R. (2010). *Rocks, jeans, and busy machines: An engineering kids storybook*. San Antonio, TX: Rivera Engineering.

Rogoff, B. (1984). Introduction: Thinking and learning in a social context. In B. Rogoff & J. Lave (Eds.), *Everyday cognition: Its development in social context* (pp. 1–8). Cambridge, MA: Harvard University Press.

Romance, N. R., & Vitale, M. R. (2001). Implementing an in-depth expanded science model in elementary schools: Multi-year findings, research issues, and policy implications. *International Journal of Science Education, 23*(4), 373–404. https://doi.org/10.1080/09500690116738.

Smolkin, L. B., McTigue, E. M., Donovan, C. A., & Coleman, J. M. (2009). Explanation in science trade books recommended for use with elementary students. *Science Education, 93*(4), 587–610. https://doi.org/10.1002/sce.20313.

Tank, K. M., Moore, T. J., & Pettis, C. (2013). *The PictureSTEM project: A curricular approach using picture books to transform STEM learning in elementary classrooms.* Scientific paper presented at the 2013 American Society of Engineering Education annual conference, Atlanta, GA. Retreived from https://peer.asee.org/22611.

Taylor, B. M., Pearson, P. D., Clark, K., & Walpole, S. (2000). Effective schools and accomplished teachers: Lessons about primary grade reading instruction in low-income schools. *The Elementary School Journal, 101,* 121–165. https://doi.org/10.1086/499662.

Thiessen, D. (Ed.). (2004). *Exploring mathematics through literature: Articles and lesson for prekindergarten through grade 8*. Reston, VA: National Council of Teachers for Mathematics.

Vygotsky, L. (1986). *Thought and language*. Cambridge, Massachusetts: MIT Press.

Watkins, J., Spencer, K., & Hammer, D. (2014). Examining young students' problem scoping in engineering design. *Journal of Pre-College Engineering Education Research, 4*(1), 43–53. https://doi.org/10.7771/10.7771/2157-9288.1082.

Yore, L. D. (2004). Why do future scientists need to study the language arts? In W. F. Saul (Ed.), *Crossing borders in literacy and science instruction: Perspectives on theory and practice* (pp. 71–94). Newark, DE: International Reading Association.

Chapter 10
Novel Engineering in Early Elementary Classrooms

Merredith Portsmore and Elissa Milto

Abstract This chapter provides another rich example of an integrated approach to early engineering education, namely the *Novel Engineering* program, which is designed to teach engineering and literacy in elementary and middle school classrooms. Through this approach, students derive engineering problems from classroom texts and then move through the engineering design process as they build solutions that are influenced by the characters, settings, and plots about which they are reading. The chapter introduces the Novel Engineering approach, shares how it has been implemented at the early elementary level, and leads the reader through examples of engineering and literacy integration. Finally, the chapter discusses what research associated with this project has found.

10.1 Novel Engineering Overview

Many professional engineers have rich contexts in which they design (Jonassen, Strobel, & Lee, 2006). They have multiple stakeholders with different needs that they must translate into design requirements; they have constraints on materials, time, or solution type they need to account for and balance; and they must address regulations and ethical issues. Professional engineers in these contexts are skilled at finding problems, identifying requirements, and balancing trade-offs. The recent calls for engineering in K-12 such as *The Framework for K-12 Science Education* (NRC, 2011) and the *Next Generation Science Standards* (NGSS, Lead States, 2013) have reinforced the importance and need to engage young students in all of these practices of engineering. Often, however, when we transpose engineering into K-12 settings, some of the richness and wonderful messiness of real-world engineering is lost in engineering activities that specify the problem and all the requirements for students. For example, the popular design challenge to build a tall tower out of spaghetti and marshmallows (e.g., Yakacki, Heavner, Zarske, & Carlso, 2004) doesn't have a client

M. Portsmore (✉) · E. Milto
Center for Engineering Education and Outreach, Tufts University, Medford, USA
e-mail: Merredith.Portsmore@tufts.edu

L. English and T. Moore (eds.), *Early Engineering Learning*, Early Mathematics Learning and Development, https://doi.org/10.1007/978-981-10-8621-2_10

and specifies the use of 20 marshmallows and pieces of spaghetti and the way the tower will be tested. This and other typical activities are great ways for students to engage in some elements of engineering design and are often necessary with the realities of school. However, they don't allow students to engage in the messiness of engineering, which includes elements like problem scoping and defining constraints. Novel Engineering works to replace real-world clients and contexts with those from literary texts in order to offer students opportunities to enter into engineering design practices that are messy, ill-defined, and without predetermined requirements or constraints.

In Novel Engineering, students use literature as the basis for engineering design challenges, drawing information from the text to identify engineering problems, considering characters as clients, and using details from the story to impose constraints as they build functional solutions in their classroom to the characters' problems. For example, students reading the book *Danny Champion of the World* by Roald Dahl identified Danny's father falling into a pit as a problem and then built and tested functional models to get his father out of the pit in a way that would use appropriate resources from the story's setting. As students work on text-based engineering projects, they also engage in productive and self-directed literacy practices, including noting key details in the text, making inferences, and writing lists and other notes that support their design process. Novel Engineering tasks are, therefore, truly interdisciplinary efforts from which students engage in both engineering and literacy activity.

Through the classrooms that have participated in Novel Engineering, there is evidence that this interdisciplinary approach is able to bridge the learning goals of engineering and literacy while meeting the Next Generation Science and Common Core Standards as well as classroom and individual student goals. A core strength of Novel Engineering is that it is not a fixed curriculum that only works with particular books but rather that it is an approach that works with a myriad of books and texts typically used within K-8 English Language Arts instruction. By leveraging required texts, teachers are able to easily integrate Novel Engineering with existing curricula and customize their approach based on their own goals and comfort level. This is particularly important at the elementary level where instructional time is a limitation and teachers often have little experience with engineering (Duncan, Diefes-dux, & Gentry, 2011). In that way, Novel Engineering allows elementary teachers to leverage their expertise and comfort with literacy and build on the engineering elements.

10.2 A Sample Novel Engineering Classroom

The following example gives an illustration of a basic elementary implementation of Novel Engineering.

Mrs. Everest, a first-grade teacher, set aside four hours spread across four days for her students to complete a Novel Engineering unit based on *Peter's Chair* by Ezra Jack Keats. A Novel Engineering unit begins when students **read a book**, independently, in small groups, or most likely, as a read-aloud in younger grades. In *Peter's Chair*, the main character, Peter, feels neglected since his mother is having a new baby and the chair he has had since he was very little will be given to the new baby. As Mrs. Everest reads the simple picture book, she stops at important parts to have class discussions so that students get a better understanding of the story, the characters, and connect to the problems that Peter is having.

Upon finishing the book, Mrs. Everest facilitates a discussion about engineering and how they can use engineering to solve a problem in the book. Class discussions move between **identifying problems** and brainstorming possible solutions that would be feasible in the book and in the classroom. Mrs. Everest's class identifies the problem of Peter feeling neglected and feels that a new chair that is *just* for Peter would make him feel better. As a class, they talk about how they will be able to know if their design works. The students and Mrs. Everest decide the chairs the students make must be able to stay together when an 18" doll is placed on them and must hold the doll at least six inches off the floor.

In an effort to **scope problems and design a solution**, pairs of students begin to discuss what their chair will look like, which materials they can use, and begin to sketch ideas. Students discuss what Peter, their client, would want for his chair. Some students think about the color, while other talk about places for his dog to sit or for him to store his toys. As the students work, Mrs. Everest walks around the room asking the pairs about their individual designs. The students use the dolls **to test and get feedback** on the size and stability of their designs. Most groups' chairs are not sturdy enough to hold the doll without falling apart so they must analyze their chairs to figure out what part is not working. The students redesign to better meet the stability criterion.

During the third day, students do **a mid-design share** about their chairs, talking about special features and problems they encountered. Mrs. Everest and the other students offer suggestions for improvement. A group that has decided to cover their chair in pink fabric is challenged by other students with evidence from the book that Peter did not like pink. During the fourth and final day, the students write letters to Peter to **share** the chair they made for him and the features that they included to meet his needs. Pairs of students have chairs built for two (to accommodate his new baby sister one day) and chairs that can be easily transported with wheels (as he ran away in the book). Through this unit, the students have read, thought and conversed deeply about the book and their designs, moved through the engineering design process, and written a letter in which they explain their design. The heart of the Novel Engineering process is that students have the opportunity to develop and to find evidence for their own designs while solving a messy, complex problem.

10.3 Novel Engineering Framework

This section details the principles and framework that supported the development and design decisions of Novel Engineering.

10.3.1 Engineering in Novel Engineering

Novel Engineering used the Tufts Center for Engineering Education and Outreach (CEEO) definition of engineering as *applying understanding of the environment in the pursuit of solutions to problems, solutions in the form of new objects, systems, or processes*. This definition emphasizes that engineering is the *pursuit* of solution to problems—an active process of solution finding that involves applying knowledge. This definition encompasses what professional engineers do as they apply formal knowledge in math and science with disciplinary engineering knowledge and rigorous engineering practices. It also encompasses what K-12 students do as they act as beginning engineers leveraging their developing knowledge of science, mathematics, materials, and their own experiences in the world.

Novel Engineering research also focused on the importance of problem scoping within engineering. Problem scoping was a focus because it represents an essential part of engineering. The problems engineers work on are typically ill-defined—meaning they have missing information, vague requirements, and multiple criteria for success (Jonassen et al., 2006). This necessitates that engineers engage in refining the problem, identifying the requirements, and defining the constraints (e.g., Cross & Cross, 1998). This practice connects professional engineers to client and problem. In Novel Engineering, engineering design begins with the identification of problems for characters and includes framing problems, conceptual planning, building and testing ideas, and sharing.

10.3.2 Literacy and Text in Novel Engineering

The word literacy is often equated with the act of reading. Some take it a step further and say that literacy is the ability to read and write, when in fact, literacy is an umbrella term that encompasses all that it means to be a literate person: reading, writing, speaking, and listening. Novel Engineering units are aligned to Common Core State Standards for English Language Arts in reading, writing, and speaking (http://www.corestandards.org/ELA-Literacy/). They leverage literacy instruction in a manner that is consistent with the belief that children can engage in complex work in engineering and literacy.

In Novel Engineering, the text serves as the basis of discourse, argumentation, and sharing of ideas and thinking. Overall, the text is a shared experience or document from which partners can work and engage in engineering design. By discussing and comprehending the text together as a group, a whole class, and with partners, students have a common language, understanding, and context. Because a text can have multiple interpretations, it makes it an ideal basis for the type of work Novel Engineering promotes—work wherein students use what they know, evidence from the world or book around them as they interpret and synthesize this information to address the problem at hand. It is in this context that we see Novel Engineering functioning well—when literacy endeavors are active and thoughtful activities through which meaning is created.

10.4 The Novel Engineering Arc

Figure 10.1 shows the general flow of a Novel Engineering unit, which was highlighted in the sample Novel Engineering classroom.

The Novel Engineering Arc comes from a combination of traditional representations of the engineering design process, researchers' analysis and observation of classroom activity, and input from collaboration with Novel Engineering educators. While there have come to be other literacy-based engineering projects (e.g., Moore & Tank, 2014; Wilson, Smith, & Householder, 2014), Novel Engineering had few resources from which to draw at its inception in 2010. The research team[1] started with models of the engineering design process for children (e.g., "Engineering is Elementary," n.d.; Massachusetts Department of Education, 2001) and looked to see

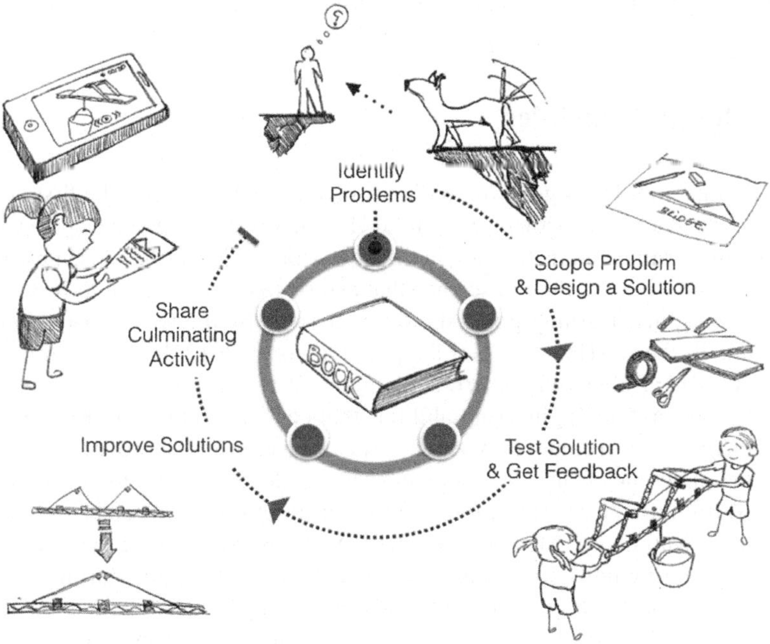

Fig. 10.1 Novel Engineering Arc

[1]The original Novel Engineering research team included Bill Wolfson and members of Engineering Lens (http://www.integratingengineering.org/) which had previously implemented a project that integrated engineering and literacy that served as the inspiration for Novel Engineering.

Table 10.1 Elements of the Novel Engineering Arc—combining engineering and literacy

Engineering design practice	Connections between engineering and literacy
Identify problems	While reading the text, students and teachers identify problems the characters face
Scope problem and design solution	Students identify which problems are related to engineering and decide what they could make to address a character's problem
Test solution and get feedback	Mid-design share-outs facilitate students getting feedback on their solution based on its functionality and how their solution would work for the character (client)
Improve solution	Students reconsider the literacy text and the character's needs as they redesign
Share culminating activity	The final presentation of the solution is done within the context of the book. (e.g., *How would the character use the solution? In what ways might it change the narrative?*)

how literature could provide characters to play the role of client. Table 10.1 illustrates how the Fig. 10.1 elements connect engineering and literacy.

10.5 Design Principles

Novel Engineering leverages two guiding principles in its design. The first principle is that teachers are professionals who should be empowered to make responsive decisions about learning and instruction in their classroom. This principle is grounded in work on responsive teaching in science that asserts that pedagogical moves, instructional choices, and learning goals should be derived by the teacher from students' action and discourse (Hammer, Goldberg, & Fargason, 2012; Levin, Hammer, Elby, & Coffey, 2013; Roberston, Scherr, & Hammer, 2015). The second principle is that children have nascent engineering abilities that can be capitalized on by providing them with meaningful contexts in which they can develop and improve their practices. This principle comes from work showing how students are able to engage in engineering design practices with minimal direct instruction in particular rich contexts (Portsmore, Watkins, & McCormick, 2012; Watkins, Spencer, & Hammer, 2014). Novel Engineering units are student-driven, and the teacher's job during a unit is to act as a facilitator, guiding discussions and engineering by listening to their students and then responding to what students are thinking and doing in an effort to help them realize their ideas and build functional solutions that respond to the problems and requirements that have emerged from the text.

Table 10.2 Novel Engineering student work and the related Next Generation Science Standards for K-2

Next Generation Science Standards		Novel Engineering Arc					
		Read book and identify problems	Scope problems and brainstorm solutions	Design solutions	Get feedback	Improve solutions	Reflect and share
K-2 Standards	K-2-ETS1-1. Ask questions, make observations, and gather information about a situation people want to change to define a simple problem that can be solved through the development of a new or improved object or tool	X	X				
	K-2-ETS1-2. Develop a simple sketch, drawing, or physical model to illustrate how the shape of an object helps it function as needed to solve a given problem		X	X		X	X
	K-2-ETS1-3. Analyze data from tests of two objects designed to solve the same problem to compare the strengths and weaknesses of how each performs				X	X	

10.6 Bidirectional Benefits and Standards

Engineering and literacy are equal components of a Novel Engineering unit and connect to both Common Core (National Governors Association Center for Best Practices & Council of Chief State School Officers, 2010) and Next Generation Science Standards (Achieve Inc., 2013).

Tables 10.2 and 10.3 list the overarching literacy and engineering components that Novel Engineering supports.

Although Novel Engineering can be connected to point to specific engineering and literacy skills and standards, the work students are doing is interdisciplinary and the students do not see or work in disciplinary silos.

Table 10.3 Novel Engineering student work and relate Common Core English Language Arts and Literacy Standards

Common Core Standards English Language Arts Literacy		Novel Engineering Arc					
		Read book and identify problems	Scope problems and brainstorm solutions	Design solutions	Get feedback	Improve solutions	Reflect and share
Reading	CCRA.R.1. Read closely to determine what the text says explicitly and to make logical inferences from it; cite specific textual evidence when writing or speaking to support conclusions drawn from the text	**X**			**X**		**X**
	CCRA.R.2. Determine central ideas or themes of a text and analyze their development; summarize the key supporting details and ideas	**X**					**X**
	CCRA.R.3. Analyze how and why individuals, events, or ideas develop and interact over the course of a text	**X**	**X**				**X**
Writing	CCRA.W.1. Write arguments to support claims in an analysis of substantive topics or texts using valid reasoning and relevant and sufficient evidence						**X**
	CCRA.W.2. Write informative/explanatory texts to examine and convey complex ideas and information clearly and accurately through the effective selection, organization, and analysis of content						**X**

(continued)

Table 10.3 (continued)

Common Core Standards English Language Arts Literacy		Novel Engineering Arc					
		Read book and identify problems	Scope problems and brainstorm solutions	Design solutions	Get feedback	Improve solutions	Reflect and share
	CCRA.W.3. Write narratives to develop real or imagined experiences or events using effective technique, well-chosen details, and well-structured event sequences						X
	CCRA.W.6. Use technology, including the Internet, to produce and publish writing and to interact and collaborate with others						X
	CCRA.W.9. Draw evidence from literary or informational texts to support analysis, reflection, and research			X	X		X
	CCRA.W.10. Write routinely over extended time frames (time for research, reflection, and revision) and shorter time frames (a single sitting or a day or two) for a range of tasks, purposes, and audiences			X	X		
Speaking and listening	CCRA.SL.1. Prepare for and participate effectively in a range of conversations and collaborations with diverse partners, building on others' ideas and expressing their own clearly and persuasively		X	X	X	X	

(continued)

Table 10.3 (continued)

Common Core Standards English Language Arts Literacy		Novel Engineering Arc					
		Read book and identify problems	Scope problems and brainstorm solutions	Design solutions	Get feedback	Improve solutions	Reflect and share
	CCRA.SL.2. Integrate and evaluate information presented in diverse media and formats, including visually, quantitatively, and orally			X			
	CCRA.SL.3. Evaluate a speaker's point of view, reasoning, and use of evidence and rhetoric			X	X	X	
	CCRA.SL.4. Present information, findings, and supporting evidence such that listeners can follow the line of reasoning and the organization, development, and style are appropriate to task, purpose, and audience		X	X	X	X	
	CCRA.SL.5. Make strategic use of digital media and visual displays of data to express information and enhance understanding of presentations						X
	CCRA.SL.6. Adapt speech to a variety of contexts and communicative tasks, demonstrating command of formal English when indicated or appropriate		X	X	X		

10.7 Elementary Classroom Engagement in Novel Engineering

The core of Novel Engineering happens when students are discussing the text with the lens of engineering. To better illustrate how Novel Engineering works, this section will share discourse from two classrooms that show students' engagement in literacy and engineering practices.

10.7.1 Mrs. Kent's Second Grade—Clementine *by Sarah Pennypacker*

Ms. Kent has taught second grade for four years and has long used *Clementine* by Sarah Pennypacker as an interactive read-aloud with her entire class. In this example, we'll see students doing several key practices from literacy and engineering: using evidence from the text to back up their thinking and ideas, connecting the fictional world to their own experience, and making inferences and predictions based on what they have read and how they think the character will respond.

After working with the Novel Engineering project, she decided to incorporate the book into a Novel Engineering unit. As the class read the book, they compiled a list of problems on an anchor chart that the characters encounter in the text. One of the problems in the book that resonates with the class is when Clementine's friend Margaret gets glue in her hair. Clementine cuts the glue-laden hair off, making her friend almost bald. This results in Margaret feeling self-conscious about how she looks. Ms. Kent stops at a point in the book when Clementine's friend is regretting having Clementine cut her hair. The teacher facilitates a discussion so that the whole class works on a conceptual design to the problem of being nearly bald. One student, Chava, thinks a hat would be a good option and is prompted by the teacher to talk about the type of hat Margaret would wear.

10.7.1.1 Using Evidence from the Text to Support Thinking and Ideas, Making Inferences and Understanding the Client

Teacher: If you have come up with at least one possible solution, show me your quiet coyotes, and let me see, okay. I heard some pretty interesting things and some ideas that I hadn't thought of.

…

Teacher: Umm, Chava?

Chava: She could wear like a pretty hat that would like; no one would notice she has, um, hair missing because it is so pretty.

Teacher: She could wear a hat. A pretty hat. Okay, so I'm going to say, "wear a hat."

Chava feels that it is important for the hat to be pretty based on Margaret's personality, and as the class talked about the characters, they noticed that Margaret likes to wear pretty things. Chava has taken this information found in the text as she thinks about possible solutions.

10.7.1.2 Making Inferences and Predictions and Balancing Trade-Offs

As the discussion progresses, Ms. Kent wants her students to consider trade-offs, which she calls pros and cons, as they think about solutions, first with a partner and then as a whole class. The discussion about trade-offs leads the students to predict what would happen if Margaret used their solution ideas. Students continually infer what her reasons would be for using these solutions in answer to questions from peers and friends. We zoom in on Cecilia and Ester's conversation as they talk about a wig as a possible solution.

Cecilia: The good thing about it is that um- the good thing about the wig is that no one will see her hair, and she'll- and she'll be able to go to school. The bad thing about it is they might um- they might like feel it or see it's fake.
Ester: Or- or go like upside down and stuff or the wig will fall off.
Cecilia: yeah
Ester: I think like a con and a pro. It would be- it's like she could have the wig and just like glue it onto her hair until the hair grows out, but the con about it would be because then- then it would never come off, so…
Cecilia: Sticking like taping it to her head.
Ester: But tape falls off real easily.

The girls present some of their ideas to the whole class.

Teacher: Okay, Ester, lead us.
Ester: Umm, me and Cecelia were thinking of pros and cons too and we-
Teacher: For the wig?
Ester: Yeah for the wig, and one of the pro was that she could still go to school and no one would really notice it. But the bad one would be, what happens if like she goes upside down or something or–
Marlena: On the monkey bars.
Cecelia: Or if, um, and wait lemme say.
Teacher: I'm going to just put a little extra star next to it because it could fall off and
Cecelia: Or if they like touch her hair they would feel that it's fake.

Celia and Ester feel that a wig would allow Margaret to feel comfortable enough to go to school, but name a few features they feel would impact Margaret negatively. They also imagine how the wig would function physically.

The teacher asks the class to discuss the trade-offs for another solution on the student-generated list—a hat.

Teacher: How about solution number two - wear a hat? Chava.
Chava: Um, um. So uh, something that's good is that she could- if they ask, "Where is all your hair?" you could say, 'I just put it behind like this.
Teacher: Okay so she could say that she, she tuck- she could say she tucked her hair under?
Student: That would be kind of like lying.
Dash: That would be a lie.
Teacher: Well, maybe that's the con. It may not- she may not feel good about it, okay? She could say her hair is tucked. Okay? How about a con about the hat? Ester?
Ester: A con about the hat is what happens if like someone wants to feel it or something, and- and or like the hat would fall off really easily.
Teacher: Okay, so could fall off.
Ester: It's really- It's really hard to do it. It's kind of like a wig 'cause if you glue it on, her hair should-
Teacher: Absolutely. Could fall off easily. It's a very similar issue as the wig.
Ester: But when the hair grows down, it- it can go past it really, but not really with the wig. But it still get- like the hair hurts a lot because all of the- the hat is glued there.
Teacher: It could be uncomfortable is what I'm getting from that. Harrison?
Harrison: A con of the hat…
Teacher: Yes
Harrison: Is um that it's- it might be sweaty while like it's gonna be on all day, so it might be sweaty.
Teacher: Sweaty.
Student: Even in the summer.
Teacher: Okay, Harry?
Harry: Well, you can't have the hat on at school.
Dash: I know! I was gonna say, you're not allowed to wear hats in school!

(Students agree)

Jack: That's not a good idea.
Student: Maybe you could tell the teacher.
Teacher: It's a possibility, but I don't know. A lot of people might be asking her questions if- even if she gets permission, so she'll be drawing-
Student: No, she can ask the principal!
Teacher: But do you think that'll stop her friends from asking her questions about why she's allowed to wear the hat? It might draw more attention to the issue. Right? It's a great point, Harry.
Ester: And people will start wearing hats and break the rules.
Teacher: Who else has a pro or a con about the hat? Alice?
Alice: Well, I sort of did- did things randomly. I didn't find any pros, but I thought-

Teacher: That's okay. That's not random.
Alice: It would have to be a really big hat for the kids to believe that Margaret could actually tuck all her hair into it.

In this exchange, the students (and teachers) are making inferences backed by information from the book and what they know about the world to imagine how a hat would make Margaret feel and how it would work as a solution to hide her baldness. The students' comments are evidence of their ability to make logical predictions based on text as well as to support them as they work on their speaking and listening skills as outlined in Common Core ELA Standards. They are also balancing multiple constraints.

10.7.1.3 Example: Making Connections

As the students are talking about the trade-offs of the wig and hat, they are applying what they have learned from their personal experiences to talk about how this would affect the function of the solutions. In this next exchange, Claire has worn a wig as part of a costume and knows they are sometimes uncomfortable.

Teacher: Absolutely, any more pros or cons about the wig? Claire?
Claire: I have a con. Sometimes wigs are really itchy…

The class continues to generate trade-offs of the wig, thinking about things they do at school such as going on the monkey bars and touching each other's hair.

Ester: Umm, me and Cecelia were thinking of pros and cons too and we–and one of the pros was that she could still go to school and no one would really notice it. But the bad one would be, what happens if like she goes upside down or something or…
Marlena: On the monkey bars.
Cecelia: Or if, um, and wait lemme say.
Teacher: I'm going to just put a little extra star next to it because it could fall off and…
Cecelia: Or if they like touch her hair they would feel that it's fake.
Dashiell: Oh yeah. That's a good one.

Bringing up the possibility of the wig fall off leads the teachers to ask the students about how they were thinking the wig would be attached. This leads to a discussion about possible materials they could use to attach it and their properties.

Teacher: What were you guys saying about how you would attach it to her head?
Ester: Well, because, well…
Cecelia: Well I was thinking tape.
Ester: But I was thinking glue but then, but then when her, when some of her hair starts to grow back then she wouldn't need the wig anymore, so it's kind of like in the middle.

Teacher: That's a great point.

Cecelia: I was thinking she could get like a clear kind of like string or something and she could tie it to her head.

Teacher: Those are some really good ideas about how, If you, if you chose the wig as a solution ways to help solve some of the issues that could arise, it falling off and such, depending on her activities. So, for you, you might be leaning more towards this solution because you're finding ways to deal with some of the cons that you're coming across.

Atticus: Umm, a con about the glue if the hair grows back under it, it would really hurt to take the wig off.

Teacher: Okay, so it could hurt to take it off.

Jack: And the hair might get stuck on the, on the, on the, on the glue and…

Ester: And it might just rip off before it even grows back.

Jack: … and yank the hair off right away.

These interactions illustrate that the teacher has identified a place in the text that leads to a discussion that begins with an open-ended question. The students are clearly comfortable engaging in open discussions as a class. This opportunity for students to discuss, listen, and share ideas in an authentic way is a core value in Novel Engineering and what makes it appealing to students and teachers. Students are given the freedom to play with ideas as they identify constraints that would influence the success of a design. They bring pertinent information from the text to help them frame the problem. Their conceptual plan begins with the consideration of multiple solutions, and their design decisions are based on evidence from the text and the world around them rather than random decision making.

10.8 Second Grade—Designing, Testing, and Evaluation

To look more at how students actually engineer and create, this section of the chapter looks at another second-grade class, Mrs. Adams' class, which is reading *Danny Champion of the World* by Roald Dahl. The class has identified a number of problems in the story, from the challenge of Danny and his father getting food to a key moment in the story when Danny's father falls in a hole. Pairs of students have selected the problem they want to focus on and are working to design, build, and test their solutions. They have some materials from the recycle bin (paper towel tubes, boxes, containers) and LEGO pieces available, and they are also allowed to request specific materials like Play-Doh, pulleys, fabric. The students are planning and building, with testing ongoing throughout the design process.

Two girls, Jeslen and Sabriel, have chosen the problem of getting the father who is injured out of a pit in the woods. They explain why they're making their solution while referring to events in the text.

Researcher: So how'd the Dad fall in the pit? I've never read this one.

Jeslyn: Because he was walking at night and there was like a hole…
Sabriel: He didn't see it, and then he fell into it.
Researcher: At night?
Jeslyn: In the woods at night. And he didn't get to see the pit so he just fell in.
Researcher: So did you read the whole book?
Jeslyn: Yeah
Researcher: So he gets out eventually, but you're thinking of another way? Or did he not get out of the pit? What ends up happening, how does he get out?
Jeslyn: Um, he got out of the pit with Danny.
Sabriel: But they didn't tell how. That's why we're doing it.
Jeslyn: Yeah, they didn't tell how.

Since they will not be able to test their solution with a real tree, they build a tree with available materials as part of a test. The tree is not yet able to stand on its own. Once it does, they will attach a pulley to a branch of the model tree they have constructed. There will be a chair connected to the pulley system to pull the father up and out of the pit (a trashcan with a Barbie doll on the bottom).

Jeslyn: This is the tree, but we're not finished, and since it couldn't stand that much, right now we're making this stand, and we're gonna put it up like that so it could stand more.
Researcher: Okay, so the dad is in the pit and how's the tree gonna help him get out?
Sabriel: 'Cause it's gonna hold up the pulley.
Jeslyn: And Danny's gonna pull up.
Sabriel: He's (Dad) gonna like get in it, and he's gonna- and Danny's gonna try to like pull him up.

…

Jeslyn: And we're gonna make a chair, so he could go up, and when his legs hurt, we're gonna make a pillow.

…

Researcher: So how are you gonna use the pulley? How's it gonna attach?
Jeslyn: We're gonna put a stick sticking out, and we might tape it on, and then we're gonna put a pulley on it, and the string- the string is gonna be in here (on the pulley), and we're gonna attach it to that.

In this exchange, the girls have planned a design and are realizing not only the devise that will get Danny's father out of the pit, but have also thought ahead to how they will test the device. Due to the scale of the pretend pit (a trashcan) they feel the need to augment the test to more closely mirror what they would need to test their design in real life. Halfway through the building time, the teacher stops the students so each group can share with the class to get feedback on their designs. Jeslyn and Sabriel explain their design and then Christian asks a question that causes the girls to think more deeply about their design.

Teacher: What's your problem?
Jeslyn: We did um, Dad falling in the pit. And this is a tree we didn't get to finish yet.
Sabriel: This is a pillow he's supposed to sit on.
Jeslyn: Um, put it leg on. And we didn't make the chair yet, and um, we're gonna tie this pulley to um this (the tree), and then um Danny's going to be pulling his dad up, and the chair's going to be tied to it. We're gonna make- Do you know how like a chair, like you just put your legs on the cushion? Yeah, this is part of it, because his leg really hurts.
Teacher: Questions or comments? … Karun.
Karun: Um, I have a comment. I think it's a good idea because in that chapter they were talking about how Danny was gonna get- was driving Baby Austin there to get, um, his dad, and since he used the rope from the car that had the pulley, I think he could just use the pulley from the car and the rope, so it does make sense to have a pulley and a rope. (Argumentation with evidence)

…

Christian: Wouldn't it be kind of heavy to pull him?
Jeslyn: Um. We're not sure about that
Sabriel: Maybe Danny might like tie it to a rock and then he might like pull that rock

Later when they return to building, Jeslyn and Sabriel make changes to their design based on the feedback that they got from the students.

10.9 Implementation Considerations for Early Elementary Novel Engineering

Evaluation of the project was collected through surveys, teacher interviews, and observations, and collated by an external, independent project evaluator as well as researchers. This evaluation yielded a number of particular considerations for Novel Engineering in early elementary classrooms.

- Books—Books that work well across Novel Engineering have rich characters and settings from which students can identify engineering problems and constraints. Across Novel Engineering, texts that have worked well have realistic setting (not magical, like Harry Potter) and problems that have multiple possible solutions. Early elementary teachers used picture books as well as longer early chapter books that they read aloud. The short nature of picture books provides some challenges as often the story or characters are simple, but are good options for initial Novel

Engineering experiences. Some short picture books that have worked well in Novel Engineering contexts include *A Snowy Day* (Ezra Jack Keats), *Peter's Chair* (Ezra Jack Keats), *Westlandia* (Paul Fleischman), and *Muncha! Muncha! Muncha!* (Candance Flemming).

- Exploration with materials—While all students need experience manipulating hands-on materials with which they will engineer, the need can be more acute in early elementary classrooms. Early elementary teachers often allowed students more exploratory time with the craft or recycled materials before starting Novel Engineering.
- Pairs of Students—As students build their solutions to a character's problems during Novel Engineering, they discuss character traits, constraints of the setting, and how to assemble the physical materials they have chosen. At the early elementary level, teachers in Novel Engineering reported that this was done best with students working in groups of two as students are just learning communication and collaboration skills.
- Tracking Problems and Ideas—There are many methods of having students track engineering problems as they read. Many classrooms used reading journals or note catching worksheets as students read independently. The methods that worked best in early elementary classrooms, where teachers most often led interactive "read-alouds," were shared visual representations, like an anchor chart (Fig. 10.2), which could be revisited each day.

10.10 Research

Research on Novel Engineering did not focus on particular aspects of the implementation in early elementary. It focused on the dynamics of how students engage with an integrated engineering and literacy approach as well as how teachers are able to navigate design problems as learners and educators, much of the data came from upper elementary. However, there are elements that have implications for the early elementary teachers.

Watkins, Spencer, and Hammer (2014) looked at how students engage in problem scoping within Novel Engineering activities, highlighting their abilities to balance criteria and to take different perspectives in a fourth-grade classroom. Similarly, Portsmore, Watkins, and McCormick (2012) examined the emergent nascent planning abilities of students when designing physical solutions for characters in a Novel Engineering unit. They found that students have rich resources for identifying materials they needed, creating representations of their ideas, and using those representations as a guide for their prototyping when the context of the project made planning necessary. Both studies raised questions about how students' engineering abilities may not be fixed but instead be context-dependent, an implication that is likely to be true for early elementary classrooms as well.

McCormick and Hammer (2016) explored the potential mechanisms for why Novel Engineering research shows students taking up engineering practices in dif-

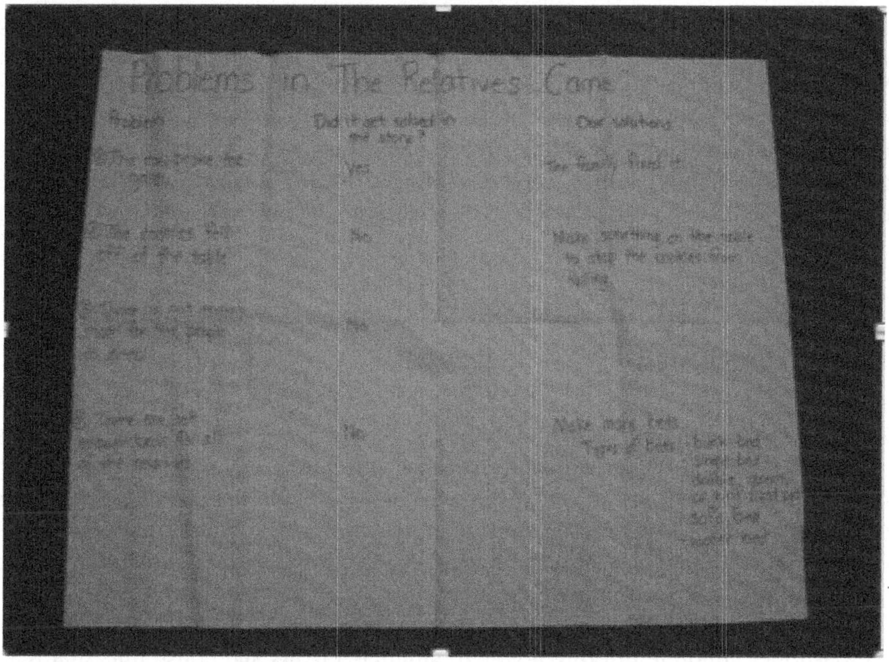

Fig. 10.2 Anchor chart from kindergarten class reading *The Relatives Came*

ferent ways. In particular, they looked at the idea of epistemological framing—how students interpret the purpose of an activity within a classroom in relation to engineering activities (McCormick & Hammer, 2016). They illustrate how students' framings shift in response to the students' sense of the purposes of the activity as well as how materials and prototypes influence students framing of the activity (McCormick & Hammer, Under Review).

Research on teachers involved in Novel Engineering has examined how teachers enter into open-ended engineering design as well as how they recognize and respond to students' engineering. Wendell (2014) looked at pre-service teachers engaged in a Novel Engineering activity and analyzed their discourse in designing a potential solution to a challenge. She found that teachers were easily able to enter into some engineering design practices, like solution generation and feasibility analysis, but did not spend time information gathering, problem scoping, or doing detailed analysis of their solution. The research suggests that teachers may need support and scaffolding with those practices.

The Novel Engineering approach's need for responsive teaching has prompted research on how K-8 teachers have taken up elements of responsive teaching and the associated challenges. McCormick, Wendell and O'Connell (2014) analyzed what teachers in clinical interviews noticed about a video of a student engaged in Novel Engineering. They identified a range of stances that teachers took toward student work. Their work laid the groundwork for thinking how responsive teaching in engineering, particularly around open-ended challenges, may develop. Johnson, Wendell and Watkins (2016) looked at clinical interviews with six teachers in which teachers discussed the challenges in identifying nascent engineering and their self-reported challenges in responding. They found that teachers, new to engineering can, with little support, notice some disciplinary practices. However, teachers were challenged in how to push students in their design when the students are struggling or straying from the problem.

10.11 Conclusion

As K-12 engineering education continues to evolve, Novel Engineering represents one of the ways in which we can emphasize different aspects of the engineering design process. Moreover, in already crowded K-8 curricula, engineering often needs to do double duty—supporting more than one discipline. Novel Engineering has demonstrated how engineering can be engaged in with disciplinary authenticity while also meeting literacy learning goals.

The context of the book has been shown to be powerful for early elementary as well as older children, offering a way to simulate clients and constraints that real-world engineers wrestle with as they define problems and identify design constraints. This authentic context has focused on ways in which students engage in engineering and the ways teachers need to be prepared to design learning situations and respond to student work.

Acknowledgements The authors would like to acknowledge all the Novel Engineering team detailed at http://www.novelengineering.org/about-us.
This material is based upon work supported by the National Science Foundation under Grant Number DRL-1020243. Any opinions, findings, and conclusions or recommendations expressed in this material are those of the author(s) and do not necessarily reflect the views of the National Science Foundation.

References

Achieve Inc. (2013). *Next Generation Science Standards* Achieve, Inc. on behalf of the twenty-six states and partners that collaborated on the NGSS.
Cross, N., & Cross, A. C. (1998). Expertise in engineering design. *Research in Engineering Design, 10*(3), 141–149. https://doi.org/10.1007/bf01607156.
Duncan, D., Diefes-dux, H., & Gentry, M. (2011). Professional development through engineering academies: An examination of elementary teachers' recognition and understanding of engineering. *Journal of Engineering Education, 100*(3), 520–539. https://doi.org/10.1002/j.2168-9830.2011.tb00025.x.
Engineering is Elementary. (n.d.). Retrieved from http://www.eie.org/.

Hammer, D., Goldberg, F., & Fargason, S. (2012). Responsive teaching and the beginnings of energy in a third grade classroom. *Review of Science, Mathematics, and ICT Education, 6*(1), 51–72.

Johnson, A. W., Wendell, K. B., & Watkins, J. (2016). *Dimensions of experienced responsive teaching in engineering*. Paper presented at the 2016 ASEE Annual Conference & Exposition, New Orleans, Louisiana. https://peer.asee.org/26846.

Jonassen, D. H., Strobel, J., & Lee, C. B. (2006). Everyday problem solving in engineering: Lessons for engineering educators. *Journal of Engineering Education, 95*(2), 139–151.

Levin, D., Hammer, D., Elby, A., & Coffey, J. (2013). *Becoming a responsive science teacher: Focusing on student thinking in secondary science*. Arlington, VA.: National Science Teachers Association.

Massachusetts Department of Education. (2001). *Science and technology/engineering curriculum framework*. Retrieved from http://www.doe.mass.edu/frameworks/scitech/1006.pdf.

McCormick, M., & Hammer, D. (2016). Stable beginnings in engineering design. *Journal of Pre-College Engineering Education Research, 6*(1), 4.

McCormick, M., & Hammer, D. (Under Review). Pretend roofs and laser beams: The role of materials and artifacts in students' framings of a Novel Engineering design experience. *International Journal of Design and Technology Education.*

McCormick, M., Kristen Bethke, W., & O'Connell, B. (2014). *Student videos as a tool for elementary teacher development in teaching engineering: what do teachers notice? (Research to practice).* Paper presented at the 2014 ASEE Annual Conference, Indianapolis, Indiana.

Moore, T. J., & Tank, K. M. (2014). *Nature-inspired design: A PictureSTEM curriculum for elementary STEM learning.* Paper presented at the 2014 Annual Meeting of the Association of Science Teacher Educators, San Antonio, TX.

National Governors Association Center for Best Practices, & Council of Chief State School Officers. (2010). *Common core state standards*. Washington, DC: National Governors Association Center for Best Practices, Council of Chief State School Officers.

National Research Council. (2011). *A Framework for K-12 science education: Practices, crosscutting concepts, and core ideas*. Washington, DC: National Academy Press.

Portsmore, M., Watkins, J., & McCormick, M. (2012). *Planning, drawing and elementary school students in an integrated engineering design and literacy activity.* Paper presented at the 2nd P-12 Engineering and Design Education Research Summit, Washington, DC.

Roberston, A. D., Scherr, R., & Hammer, D. (Eds.). (2015). *Responsive teaching in science and mathematics*. New York, NY: Routledge.

Watkins, J., Spencer, K., & Hammer, D. (2014). Examining young students' problem scoping in engineering design. *Journal of Pre-College Engineering Education Research (J-PEER), 4*(1), 5. http://dx.doi.org/10.7771/2157-9288.1082.

Wendell, K. B. (2014). Design practices of preservice elementary teachers in an integrated engineering and literature experience. *Journal of Pre-College Engineering Education Research (J-PEER), 4*(2), 4.

Wilson, A. A., Smith, E., & Householder, D. L. (2014). Using disciplinary literacies to enhance adolescents' engineering design activity. *Journal of Adolescent & Adult Literacy, 57*(8), 676–686. https://doi.org/10.1002/jaal.302.

Yakacki, C., Heavner, B., Zarske, M. S., & Carlso, D. (2004). *Leaning tower of pasta*. Retrieved from https://www.teachengineering.org/activities/view/cub_mechanics_lesson10_activity1.

Chapter 11
Books, Butterflies, and 'Bots: Integrating Engineering and Robotics into Early Childhood Curricula

Mollie Elkin, Amanda Sullivan and Marina Umaschi Bers

Abstract Although we are surrounded by technology on a daily basis, the inner working of devices like phones and computers is often a mystery to children and adults alike. Robotics offers a unique way for children (and grown-ups!) to explore sensors, motors, circuit boards, and other electronic components together from the inside out. This chapter describes how robotics can be used as a playful medium in early childhood classrooms to learn foundational engineering and computer science concepts. By presenting vignettes from three early childhood classrooms that embarked on an eight-week KIBO robotics curriculum, this chapter highlights how educators with little to no prior engineering experience were able to successfully integrate robotics with traditional early childhood content such as literacy and science. KIBO is a developmentally appropriate robotics kit specifically designed for children ages 4–7 that is controlled with tangible programming blocks—no screen time required. The three classroom teachers worked with researchers from Tufts University and Lesley University to integrate KIBO robotics with the teachers' traditional learning units. The three vignettes will describe the following classroom experiences: using robotics to bring to life the book *Brown Bear, Brown Bear, What Do you See?* in the context of literacy explorations; and in science, programming the life cycles of the frog and the butterfly, and using robots to model the movement of worms through different environments. These vignettes will highlight the very different approaches teachers took to introducing robotics to their students and how they utilized the engineering design process as a teaching tool that can be applied to most subject areas.

M. Elkin (✉) · A. Sullivan · M. U. Bers
Tufts University, Medford, MA, USA
e-mail: Mollie.elkin@tufts.edu

A. Sullivan
e-mail: Amanda.sullivan@tufts.edu

M. U. Bers
e-mail: Marina.bers@tufts.edu

L. English and T. Moore (eds.), *Early Engineering Learning*, Early Mathematics Learning and Development, https://doi.org/10.1007/978-981-10-8621-2_11

Although we are surrounded by technology on a daily basis, the inner working of devices like phones and computers is often a mystery to children and adults alike. Robotics offers a unique way for children (and grown-ups!) to explore sensors, motors, circuit boards, and other electronic components together from the inside out. This chapter provides another perspective on applying the engineering design process in integrating early engineering education within the classroom. With a focus on technology, this chapter describes how robotics can be used as a playful medium in early childhood classrooms to learn foundational engineering and computer science concepts. By presenting vignettes from three early childhood classrooms that embarked on an eight-week KIBO robotics curriculum, we highlight how educators with little to no prior engineering experience were able to successfully integrate robotics with traditional early childhood content such as literacy and science. KIBO is a developmentally appropriate robotics kit specifically designed for children ages 4–7 that is controlled with tangible programming blocks—no screen time required. The three classroom teachers worked with researchers from Tufts University and Lesley University to integrate KIBO robotics with the teachers' traditional learning units. The three vignettes describe the following classroom experiences: using robotics to bring to life the book *Brown Bear, Brown Bear, What Do you See?* in the context of literacy explorations; and in science, programming the life cycles of the frog and the butterfly, and using robots to model the movement of worms through different environments. These vignettes highlight the very different approaches teachers took to introducing robotics to their students and how they utilized the engineering design process as a teaching tool that can be applied to most subject areas.

11.1 Introduction

Anyone who has spent time with a four- or five-year-old child has undoubtedly been asked the famous "why?" questions about the world around her. *Why is the sky blue? Why do birds fly, but not dogs?* As their environment becomes increasingly technological, children's questions are beginning to include things like "how does a phone work?" and "why do some doors open automatically?" These questions are a genuine attempt for children to make sense of their world and understand how things work. This natural inclination to curiosity, inquiry, and investigation is not only the cornerstone of early childhood development, but is also a key component of thinking like an engineer (Brophy, Klein, Portsmore, & Rogers, 2008; Peel & Prinsloo, 2001).

When introducing the engineering design process to young children, we can begin by satisfying their curiosity through asking questions or posing problems that children are personally interested in investigating (see Fig. 11.1). As the chapters in this book have documented, early childhood is the ideal time to begin teaching engineering concepts because children are naturally inquisitive about the world around them and are motivated to explore, build, and discover answers to their big questions. Educators and researchers are thus beginning to see the importance of teaching engineering at an early age (Bers, 2008, 2018). According to the Massachusetts Department of

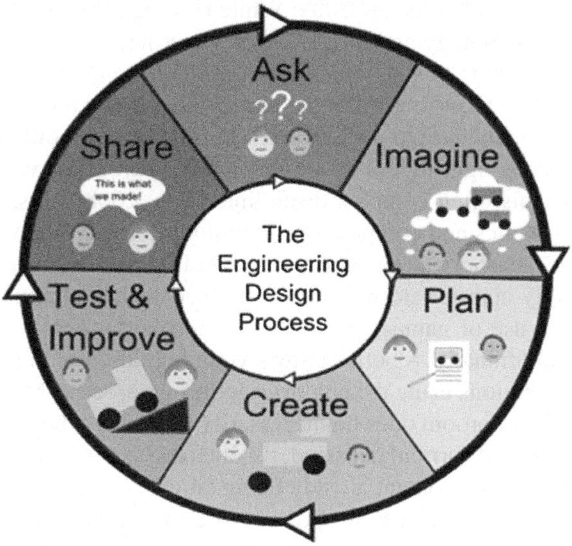

Fig. 11.1 An illustration of the engineering design process (image created by the DevTech Research Group at Tufts University)

Education (2006), the engineering design process refers to the cyclical or iterative process engineers use to design an artifact in order to meet a need. In line with the other descriptions of engineering design presented in this book, the Massachusetts curriculum frameworks refer to identifying a problem, looking for ideas for solutions and choosing one, developing a prototype, testing, improving, and sharing solutions with others. The steps of testing and improving, which require problem-solving and perseverance, are critical for establishing a learning environment where experiencing failure, as opposed to instant success, is necessary for learning (Bers, Flannery, Kazakoff, & Sullivan, 2014; Chap. 9). This is aligned with Dweck's (2006) concept of "growth mind-set." Growth mind-set refers to the belief that basic abilities can be developed through dedication and hard work. By developing this type of attitude through activities like engineering, children are improving their skills for effectively facing challenging situations. (Dweck, 2006). Growth mind-set complements the engineering design process, but it is also applicable to other areas of personal and cognitive development such as dealing with interpersonal conflicts and persevering through challenging coursework.

Explicitly teaching these foundational engineering concepts has only recently become an interest to early childhood educators. Science curricula in early childhood classrooms were traditionally more likely to focus on the natural world including plants, animals, and the weather (Bers, 2008). While learning about the natural world is important, developing children's knowledge of the human-made world, the world of technology and engineering, is also needed for children to understand the environment in which they live (Bers, 2008; Sullivan & Bers, 2015). Research and policy changes over the past five years have brought about a newfound focus on STEM (science, technology, engineering, and math) education for young children

(Sesame Workshop, 2009; White House, 2011), with particular emphasis on the "T" of technology and the "E" of engineering.

Amidst this national focus on STEM, and engineering in particular, early childhood educators are now faced with the difficult issue of *how* to implement engineering curricula in their classrooms. One of the major difficulties early childhood teachers face is figuring out developmentally appropriate ways to introduce young children to this often complex discipline (Bers, 2008; Bers, Seddighin, & Sullivan, 2013). Robotics and computer programming initiatives have grown in popularity as a way for teachers to introduce young children to engineering content in a developmentally appropriate way that is aligned with traditional teaching approaches such as the use of games, group work, and playful exploration (Bers, 2008, 2012, 2018). Additionally, robotics allows young children to build, create, and design their own inventions using the engineering design process. Moreover, integrating robotics into the classroom does not necessarily require teachers to take time away from teaching standard curricula; instead, robotics can serve as another entry point for their students to explore content already being taught.

In this chapter, we present three vignettes from a public school in Cambridge, Massachusetts, that recently began a robotics and programming initiative in their early grades (kindergarten through second grade):

(1) integrating robotics and literacy to bring to life the book *Brown Bear, Brown Bear, What Do you See?*
(2) using robotics to program the life cycles of the frog and the butterfly, and
(3) using robots to model the movement of worms through different environments.

These vignettes were chosen to illustrate how robotics can be used to facilitate the learning of foundational engineering content while being integrated into literacy and natural science curricular content. Implications for best practices in the early childhood classroom are discussed.

11.2 Robotics in Early Childhood Education

Robotics and computer programming initiatives are growing in popularity among early childhood researchers and educators (Bers, 2008; Bers et al., 2013; Elkin, Sullivan, & Bers, 2014; Kazakoff & Bers, 2014; Strawhacker & Bers, 2014; Sullivan & Bers, 2015). Recent work has shown how the field of robotics offers a unique potential for early childhood classrooms by facilitating cognitive as well as fine motor and social development (Bers et al., 2013; Lee, Sullivan, & Bers, 2013). For example, robotics can support a range of cognitive skills, including number sense, language skills, and visual memory (Clements, 1999a). New educational robotic construction sets may help children develop a stronger understanding of mathematical concepts such as number, size, and shape in much the same way that traditional materials like pattern blocks, beads, and balls do (Resnick et al., 1998; Brosterman, 1997). Technology can also serve as catalysts for positive social interactions and emotional growth in children (Clements, 1999b). For example, robotics offers a playful way for young children to practice social skills by sparking collaboration and teamwork (Bers, 2008; Lee et al., 2013). Robotic manipulatives invite children to participate in peer-to-peer interactions and negotiations while playing and learning in a creative context (Resnick, 2003).

Robotics engages young children as engineers by allowing them to construct and design with electronic and non-electronic components. It also inspires children to become storytellers by creating and sharing personally meaningful projects that react in response to their environment (Bers, 2008). The discipline of robotics provides opportunities for young children to learn about mechanics, sensors, motors, programming, and the digital domain (Bers, 2010; Sullivan & Bers, 2015; Strawhacker & Bers, 2014). The use of educational robotic kits is now becoming widespread in elementary schools (Elkin et al., 2014; Kazakoff, Sullivan, & Bers, 2013; Rogers, Wendell, & Foster, 2010; Sullivan & Bers, 2015).

Research with programmable robotics in early childhood settings has shown that beginning at age 4, children can learn fundamental programming concepts of sequencing, logical ordering, cause and effect relationships, and engineering design skills (Bers, 2008; Fessakis, Gouli, & Mavroudi, 2013; Kazakoff et al., 2013). When children create programs to make their robots move, they are sequencing commands for their robot to act out. The act of sequencing is foundational for early math, literacy, and planning (Zelazo, Carter, Reznick, & Frye, 1997). Additionally, educational robotic programs, when based on research, child development theory, and developmentally appropriate practices, can foster student learning of engineering such as design skills and methods while engaging in collaboration and other social skills necessary for school success (Clements, 1999a, 1999b; Druin & Hendler, 2000; Svensson, 2000; Lee et al., 2013).

11.3 The KIBO Robotics Kit

The vignettes presented in this chapter utilize the KIBO robotic kit (see Fig. 11.2) created by the DevTech Research Group at Tufts University after years of research funded by the National Science Foundation and now made commercially available by KinderLab Robotics (www.kinderlabrobotics.com). KIBO is designed for children ages 4–7 and consists of both hardware (robotic parts to assemble) and software (tangible programming blocks to make KIBO move). Using KIBO, children are engaged with the engineering design process while they build a functional and mobile robot using wheels, motors, lights, and a variety of sensors. These sensors, intentionally designed to resemble body parts or objects that children are familiar with, include sound (shaped like an ear), light (shaped like an eye), and distance (shaped like a telescope). Additionally, there is a light output module which resembles a lightbulb.

Unlike other programming applications and games for children that require the use of iPads and computers, KIBO is programmed to move using interlocking wooden programming blocks. These wooden blocks contain no embedded electronics or

Fig. 11.2 KIBO robotics kit

Fig. 11.3 Programming wooden blocks for the KIBO robotics kit

digital components; it is aligned with American Academy of Pediatrics' (2003) recommendations for limited screen time for young children (Sullivan, Elkin, & Bers, 2015). The robot itself has an embedded scanner that reads the barcodes on each programming block and instantly sends the program to the robot. Similar to other programming languages, KIBO has specific syntax rules to follow. For example, every program must start with a Begin block and finish with an End block. Additionally, in order to create a functional repeat loop (which makes KIBO do actions a certain number of times), one must use the Repeat block, a parameter (either a number or sensor), and the End Repeat block (Fig. 11.3).

In addition to teaching engineering and programming concepts, the KIBO robotics kit encourages creativity and artistic design in young users. The kit contains two art platforms that can be used to personalize robotic creations with arts, crafts, and

Fig. 11.4 Sample decorated art platforms on KIBO robots

recycled materials (see Fig. 11.4). The kit also inspires collaboration and teamwork. KIBO's programming blocks are tangible so they can be shared easily between multiple children who are collaborating on programming together.

Most importantly, KIBO is fun and easy to use by young learners and adults with little to no technical experience. Children can use the kit to create delightful and silly creations that dance, light up, and make noises. Unlike other toys, KIBO looks and behaves differently every time because children can alter KIBO's aesthetic appearance with craft materials, change the assembly of motors and sensors, and alter the robot's actions through new programming commands. The following vignettes illustrate the diversity of identities KIBO can take on from frogs and butterflies to live action versions of popular children's books.

11.4 School Background

The three vignettes described in this chapter took place at an urban, public school in Cambridge, MA, serving students in prekindergarten through fifth grade. The Massachusetts Department of Education reports that at the time of the curricula, the student population at the school was 32.2% White, 22.9% African American, 20.9% Hispanic, 18.2% Asian, and 5.8% Multi-Race, Non-Hispanic. English was not their first language for over a third of the students (Massachusetts Department of Elementary and Secondary Education Enrollment Data, 2015). The students came from three early childhood classrooms (one kindergarten, one first grade, and one second grade). Neither the students nor the teachers had been previously exposed to the KIBO robotics kit.

A relatively new makerspace had just been built within the school thanks to a technology partnership with Lesley University. The makerspace was created as a way to enhance teaching and learning through technology integration. As part of this initiative, the school had acquired a variety of early childhood appropriate technologies such as BeeBot, iPads, and KIBO for students to explore engineering, programming, and robotics. With this newfound abundance of technological tools available, teachers actively looked for ways to incorporate these technologies as part of their standard curriculum, rather than using them as an "add-on" to the already busy school day. With the help of Lesley staff, they decided to use KIBO as part of pilot curricula in three classrooms before rolling it out throughout all of their lower elementary classrooms. This was an ideal opportunity to try out different strategies and see what worked in the different classrooms.

11.5 Curricula Overview

The curricula presented in this chapter were created collaboratively between the three classroom teachers, the librarian, the art teacher, and researchers at the DevTech Research Group and Lesley University, leveraging each group's expertise. All agreed on three objectives for the curricula. First, the curricula needed to address fundamental engineering, robotics, and programming concepts. This would be accomplished through a variety of small engineering and programming challenges, as well as playing fun games that reinforced the concepts. Second, for the final project component of the curricula, the KIBO content needed to connect to a topic that students were already studying in their classrooms. This could be anything from science and math to literacy, but it would be classroom-specific and determined by the classroom teacher. Third, a component of each class' final projects needed to include the visual arts. Using these criteria, three KIBO curricula were created, tailored to each of the three classrooms.

The curricula were divided into eight one-hour sessions over the course of two months, each taking place in the school's makerspace. The sessions were taught by Tufts University researchers and supported by classroom teachers, Lesley researchers, and the specialist teachers (art and librarian). The first six sessions were devoted to familiarizing the students with engineering and the KIBO robotics kit. Children started by learning about the definition of a robot and an engineer through two physical games, "Jump for Robots" and "Jump for Engineers," where children jumped if they thought they were shown a picture of a robot or something an engineer made, respectively. This led to discussion about how to identify robots and human-engineered creations. To learn about KIBO's different programming blocks, they played another game called "KIBO Simon Says," where children followed the directions on large print-out versions of the KIBO blocks. As children learned more complex syntax to the KIBO programming language, the game became increasingly more challenging with more ways for Simon (the instructor) to trick them. An important topic of the curricula was the engineering design process (see Fig. 11.1), which was taught through a song and referenced during each lesson. Finally, children participated in a sensor walk around the school in order to learn about the difference between senses and sensors, as well as about each of KIBO's different sensors.

For a portion of each session, students participated in a specific engineering or programming challenge in order to practice the concept that had just been introduced. For example, during Session 1, children had learned about the engineering design process and how to put together KIBO. Their challenge that day was to assemble a sturdy KIBO robot using motors and wheels along with non-robotic art decorations that would not fall off when KIBO was programmed to shake vigorously. Children returned to the iterative design process to "test and improve" if their decorations fell off or if their motors were not attached properly. Another challenge, during Session 5, was to program KIBO to move along different shaped maps using the Repeat and End Repeat blocks. Children had just learned how to make syntactically correct programs with repeats, so their challenge was to create programs that would make KIBO travel in a straight line, in an L-shape, and in a square using these new blocks to simplify their code.

At the end of each session, time was always allotted for the sharing aspect of the engineering design process. This gave students an opportunity to present what they had created and get feedback from their peers, to discuss what they thought was easy or challenging that day, and to ask questions. Teachers could also use this as a time to informally assess which concepts their students understood and which needed more review. For example, if many children thought using the repeat blocks was difficult and many projects did not have functional repeat loops, this would be a concept that teachers knew they needed to further address. They could either review concepts during this share session itself, or return to it at the beginning of the next class through games and teacher-led demonstrations.

Students worked on their final projects during the last two sessions of the curriculum. The project chosen in each class was unique and based on unique and based on what children were already learning in the classroom. At the beginning of the curricula, teachers had not planned out their classes' final projects. They wanted to get started to see the capabilities of KIBO and how their students used the robot. Each teacher brainstormed a variety of ideas, some which would have been too complicated and others which would have been too simple, and then worked with the Tufts and Lesley University researchers to refine their ideas. During this time, the teachers were learning first-hand about how to use and apply the engineering design process.

The following vignettes describe the experiences of the teachers and students during the two sessions they spent creating their final KIBO projects. The process for creating final projects was similar in each of the classes. First, teachers reviewed the subject content (either the natural sciences or literacy) outside of the allotted robotics time. Then, students were divided into groups of two or three and they brainstormed project ideas that could be brought to life with KIBO (the "planning" phase of the engineering design process). Next, they recorded their ideas in their Engineering Design Journals. They then created their programs for their robot, tested them out, and modified them. Finally, they created artistic decorations for their robots using art, crafts, and recycled materials. All of this hard work culminated with a final presentation of their projects to classmates, teachers, and researchers at the end of the last session.

Fig. 11.5 One group's challenge to get their KIBO from the black sheep to the goldfish

Vignette #1: Brown Bear, Brown Bear, What Do You See?

An important part of the kindergarten teacher's daily routine was reading stories aloud to her students. At the time of the KIBO curriculum, her students had been reading the well-known rhyming book *Brown Bear, Brown Bear, What Do you See?* by Eric Carle. When she was brainstorming final project ideas for the KIBO curriculum, she was inspired by this book which serves as a milestone for many children's lives as emerging readers. She saw the final KIBO projects as an ideal opportunity to integrate literacy, engineering, programming, and robotics.

The biggest challenge this teacher described facing was how to use the robots with this story in a meaningful way. She wanted the project to be structured so that the book could be read along with the kids' final presentations. Additionally, she realized that the structure of the story did not lend itself to much action (which is typically a key component of robotics projects). Each page of the book presents an animal that is asked the question "What do you see?" and it responds that it sees another animal or object. After consulting with the art teacher as well as Tufts and Lesley University researchers, she came up with a plan. Students would be divided into groups and assigned one page of the story; their goal was to program their robot to travel between two pictures on the ground, with each picture representing a character in the story. For example, one group would be given the challenge of programming pages 15 and 16 of the book, so that the KIBO robot would travel from the black sheep to the goldfish (see Fig. 11.5). The pictures of the characters would be set up around the room in the order they appear in the story. Once students successfully programmed their robots to travel from one picture to another, they would be able to add additional actions for KIBO to do to bring their characters to life.

Children used the engineering design process, particularly the stages of testing and improving, as a guide when creating their programs for KIBO. First, children needed to calculate how many Forward blocks they would need to get their robot from one picture to another. This required a period of estimation and trial and error with the robots, which was at time frustrating for the students. The teacher was also challenged with providing the "right" kind of help for her students without simply telling them "it takes 4 Forward blocks." Instead, she scaffolded their learning experience by helping them measure and estimate the distance between each picture using the floor tiles as a visual guide. Eventually, all groups persevered, and as a class, they determined the correct number of Forwards between each picture.

Once students completed basic programs, their teacher prompted them to edit their programs by experimenting with repeat loops instead of using multiple Forward blocks to create a more streamlined program. Children worked to create syntactically correct programs with the repeat loop blocks and number parameters. These complex blocks allowed children to create a more concise program for KIBO only using only one Forward block. Once children successfully programmed their robots to travel from one picture to another and recorded it in their Engineering Design Journals, they had the option to add additional action blocks. Some blocks, such as Turn Left and Turn Right, would make the KIBO robot travel off course, so the kindergarteners needed to experiment with how to ensure KIBO reached its final destination. Additionally, children were prompted to consider how the different animals they were representing might move in order to capture these motions using KIBO's programming blocks. It took multiple iterations, as well as some adult support, for many of the groups to get their robots to move from one picture to the other and capture the essence of the animals from the story.

Although only one hour each week was devoted to working on KIBO robotics in the school's makerspace, this kindergarten teacher worked closely with the art teacher and used non-robotics time during the regular school day for her students to work on different components of their final projects. For example, during art class, children created decorations for their robots. Using tin foil, pom poms, colored paper, pipe cleaners, cups, and other recyclables, children worked in their groups to create sculptures that would sit on top of their KIBO robots. Additionally, during normal class time, the teacher read the book aloud several times in order to familiarize students with the characters and the order in which they appear. This also reinforced a core concept behind both programming and writing: Order and syntax impact the way a story or product is conveyed. Overall, by using non-robotics time, the kindergarten students had additional time during the last two sessions to explore the different programming instructions, plan and test their programs, and document their creations in their journals (Figs. 11.6 and 11.7).

Students presented their final projects to one another and visiting school administrators at the end of the last session. The kindergarten teacher read *Brown Bear, Brown Bear, What Do you See?* aloud as part of the presentation. After reading a page aloud, the corresponding group presented their project. For example, the group who programmed their robot to travel from the black sheep to the goldfish programmed their robot to start moving when it heard a clap (using the sound sensor); then, it moved forward three times, turned its light on, turned right, and stopped. Another group took a more direct approach for their robot to move from the brown bear to the red bird. They created a program where the KIBO robot repeated the Forward block four times and then stopped. Students and teachers expressed having a great time during the final presentations; it served as a celebration of the students' hard work over the course of the eight-week curriculum. In addition to celebrating the

Fig. 11.6 Kindergarten students creating their decorations for their robots

Fig. 11.7 Kindergarten students creating their decorations for their robots

final products, students and teachers had an opportunity to discuss their learning processes and challenges they faced along the way. This provided a meaningful way for children to express their knowledge and expertise of engineering and programming, as well as mastery of the story, as they demonstrated their robots and programs for visitors (Fig. 11.8).

Vignette #2: Life Cycles of the Frog and Butterfly

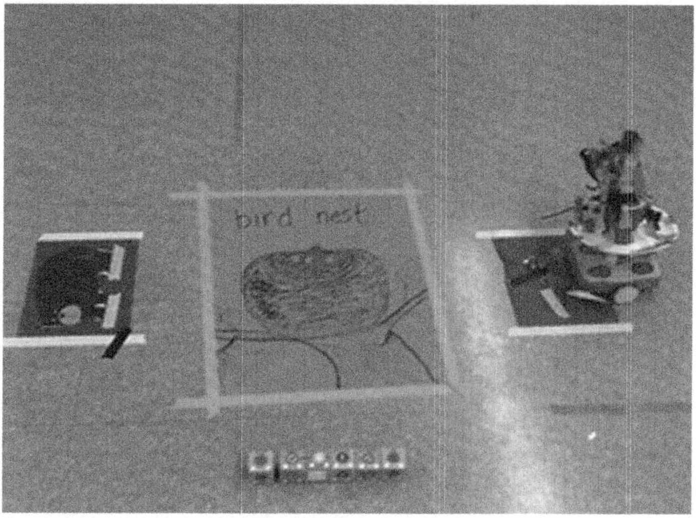

Fig. 11.8 Example final project and robot's program

At the time that the first-grade students were participating in the KIBO curriculum, they were also studying various plant and animal life cycles in their classroom. Their teacher wanted to find a way to bridge together science and robotics, so she had the idea of using KIBO to model animal life cycles. She selected the frog and butterfly life cycles because she felt they could be well represented using KIBO's different action blocks like Shake and Spin. Students were put into groups of two or three and assigned one part of a life cycle. They then were given a two-part task. The first challenge was to program KIBO to perform an action to represent movement during that stage. The second challenge was to program KIBO to move to the next stage of the cycle.

Initially, the teacher was puzzled about how to structure the final project. On the one hand, she wanted students to demonstrate through the programs they created that they understood *all* steps of one of the life cycles. Realistically, she realized it would be difficult to ask students to create four separate programs to represent each part of the cycle due to time constraints. After brainstorming with the librarian, as well as the Tufts and Lesley University researchers, she decided to have four groups recreate one life cycle, with each group focusing one part. For the frog cycle, students would be in the following groups: eggs, tadpole, froglet, and adult frog. For the butterfly cycle, students would be in the following groups: egg, caterpillar, chrysalis, and butterfly. With this idea to have four groups working on a different stage of one cycle, the first-grade students had a unique but feasible challenge: Unlike the projects in the other classes, they would need to coordinate their robots' movements with one another. This would provide ample opportunities for children to utilize the engineering design process as well as practice collaboration.

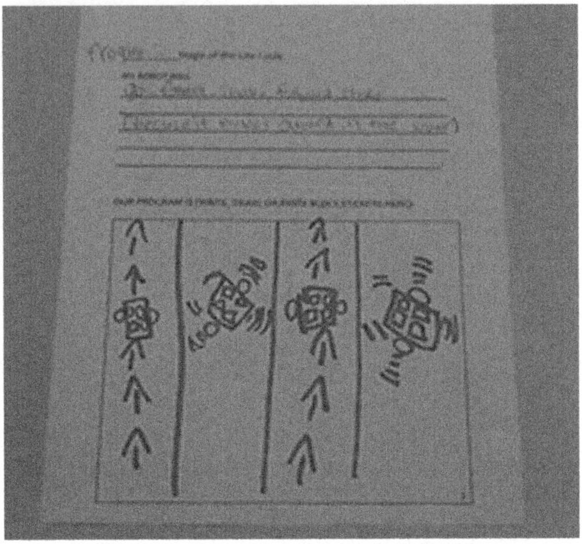

Fig. 11.9 Engineering Design Journal entries

After being assigned to their groups, children spent time reviewing their assigned cycles by watching short videos. The teacher encouraged the students to pay particular attention to the movements they saw at each part of the cycle and consider which of KIBO's programming instructions might be able to represent these movements. For example, the group working on the "frog eggs" noticed that the movement of the eggs in the water resembled how KIBO moves when it is programmed to shake. Afterward, students used their Engineering Design Journals (Figs. 11.9 and 11.10) to plan out their initial programs. The journals were designed so that children could demonstrate their understanding of the life cycle because they needed to write down what happened during their part with words, as well as illustrate the programming blocks or actions that would be used. Since KIBO does not have programming blocks for actions such as jump or fly, children had to creatively decide which blocks they wanted to use to represent these actions. For example, the group working on the froglet part of the frog cycle chose to create the program Begin, Forward, Shake, Forward, Shake, End. One child explained that this program was appropriate "because [froglets] moves around in the water."

Once each group had finished creating their programs, the teacher provided materials for students to decorate their KIBO robots. Unlike the kindergarten class, this was done during robotics time. Each group was given a printed image of how their amphibian/insect looked at their stage of the cycle; they could look at the image for inspiration or incorporate it as part of their decorations. In addition to this, children could use modeling clay, markers, paper plates, and other craft materials. Many groups faced an engineering challenge when it came to figuring out how to attach their creations to their robots. One group had the idea of placing their decorations on a plate and then attaching the plate to KIBO's art platform. After some trial and

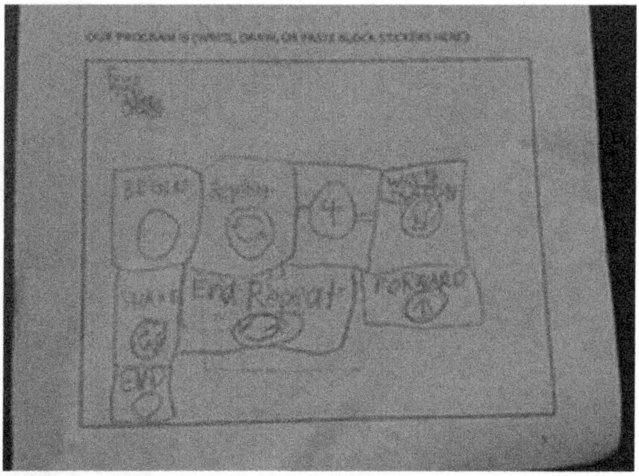

Fig. 11.10 Engineering Design Journal entries

Fig. 11.11 Decorations for the "froglet" robot

error, this group figured out how to keep the plate from falling off the platform and was able to share their idea with peers and teachers. Soon after, the other groups followed their lead. By the end, each group had successfully integrated the visual arts into their final projects (see Figs. 11.11 and 11.12 for sample projects).

Before presenting to the whole class, the four "frog cycle" and four "butterfly cycle" groups had time to practice together, making sure that each robot traveled the correct distance to reach the next part of the cycle. This process took up a substantial portion of the final session, as some groups had miscalculated the distance their robot should travel, while other groups had their decorations fall off when KIBO executed its program. The students helped one another and reinforced the idea that each group's individual programs needed to work in order to accomplish their larger goal. With time to revise and guidance from the adults, each group was ready to share their creations by the end of the last session.

Fig. 11.12 Decorations for the "butterfly" robot

During the final presentations, each group started by explaining what happened during their part of the life cycle. Next, they shared their writing and drawings in their Engineering Design Journals in order to show which blocks they used for their program and how their program represents their respective part. Finally, they demonstrated their program using the robot. After each of the four groups in a cycle had presented, there was time for students to express what they found easy and what they found challenging. Children were given the chance to showcase not only their newfound knowledge of programming and robots, but also their expertise on each aspect of the life cycles. Additionally, children demonstrated their collaborative spirit by working toward the common goal of creating one life cycle represented with multiple KIBOs.

Vignette #3: Worms and Their Environments

The second-grade students had been exploring how worms move through different environments while participating in the KIBO curriculum. As part of their final projects, the teacher wanted to connect KIBO to their unit on worm movement; she posed the following question to her students: How does terrain affect a worm's movement? Students would have to use the knowledge that they had learned in class, their programming knowledge of KIBO, and their creativity to explore this question and depict how a worm's movements changed when traveling through sand, leaves, and rocks. KIBO, acting as the worm, would need to travel along a straight line through at least one of the environments.

Children first spent time reviewing and collecting new research on the characteristics of the three different environments during science time. The teacher suggested that students use their arms and hands to model how a worm generally moves, and then try to adjust that movement based on its setting. She also provided videos and diagrams as alternative options to understanding how worms and their environment interact. Finally, she led a discussion about which terrains would make it easier or

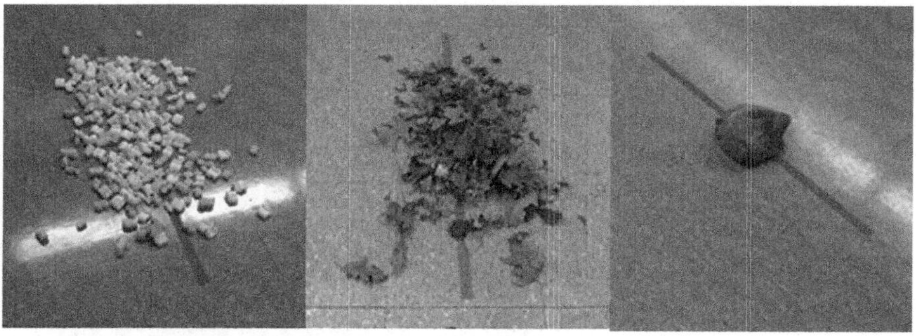

Fig. 11.13 Different worm environments

harder for worms to move. Based on their research, the class concluded that leaves would be the easiest, then sand, and then rock.

Because children were working with technology that could be damaged by sand and rocks, the environments would need to be modeled using materials that would not harm the KIBO robots. An important discussion naturally emerged about what materials could and could not be used around the robots. The children talked about how KIBO's wheels might get stuck in real sand, so the teacher and researchers provided shredded packing peanuts to be used in its place. Rather than KIBO having to drive over many different rocks, which the children hypothesized would possibly break KIBO's wheels, a large rock was used to represent the rock environment; children would have to program their robots to recognize the rock using one of KIBO's sensors and move around it. Finally, the students discovered that leaves would not harm the robots, so the teacher collected real leaves and brought them in for the projects. Before taking out the robots, children spent time examining the objects in each environment in order to help them plan out better programs. Then, the environments were set up around the room so children could reference them while they worked on their robots in their groups (Fig. 11.13).

Both children and adults needed to use the engineering design process as they created their final projects. For example, the sand environment was represented with the packing peanuts. When children programmed KIBO to travel forward, the robot would not always go straight because the packing peanuts were slippery and their size obstructed the motion of KIBO. The teacher herself was an engineer as she rethought the way the materials would be best used. After trying multiple solutions and brainstorming with researchers, she decided to cut the packing peanuts into much smaller pieces so that KIBO could move more easily through this environment. By doing this, adults modeled the iterative process of engineering and how to problem solve through a frustrating situation. Additionally, they demonstrated an "everyday" application of the engineering design process.

Once students finished creating and testing their programs, their next challenge was to incorporate the visual arts. With modeling clay and other craft materials, they created models of worms (see Figs. 11.14 and 11.15). However, they were not sure

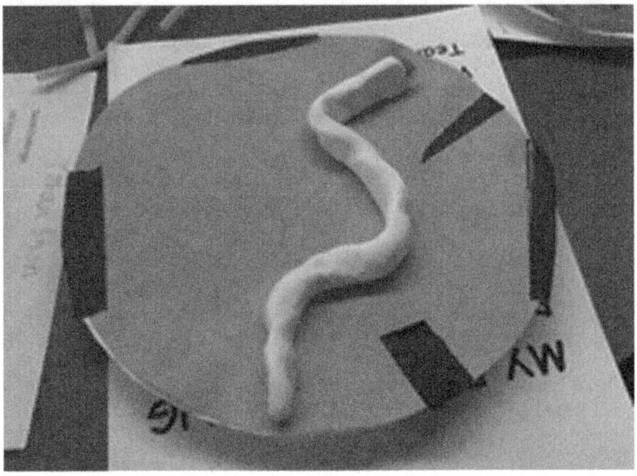

Fig. 11.14 Decorations for the robots

Fig. 11.15 Decorations for the robots

at first how to securely attach their worms to the robots. The teachers took this as an opportunity to demonstrate that engineers often borrow and improve on each other's ideas by sharing the first-grade class' creations and suggesting a similar method of attaching their worms to a plate. This worked well, but students still needed to troubleshoot strategies so that their decorations would not fall off when the robots were in motion.

For their final presentations, each group demonstrated their engineering, programming, and science knowledge as they shared their programs for one of the environments. As they shared, students described their robot's movements and why they were unique for that particular terrain. At the end, students had time to discuss the similarities and differences between the groups' programs based on the environment their robot was traveling through. This was a very unique curriculum experience because students guided much of their own learning. From spontaneously testing out sensors to deeper discussions of the robotic elements in KIBO, the second-grade

Fig. 11.16 Second-grade students presenting their final projects

Fig. 11.17 Second-grade students presenting their final projects

students took their teacher's plan in a personally meaningful direction based on their own curiosity (Figs. 11.16 and 11.17).

11.6 Discussion

These three vignettes highlight the iterative process of creating and implementing a robotics curriculum to not only teach about foundational engineering content, but also integrate literacy and natural science curricular content. For example, the first-grade students needed to draw on their scientific knowledge of how their animal moved during the life cycle before they could effectively represent this with a program.

Similarly, the kindergarten students needed to be familiar with the sequence and story line of the *Brown Bear, Brown Bear, What do you See?* book in order to bring this story to life with robotics. The final projects exemplify the diversity of creations and the integration of robotics with traditional early childhood content.

11.7 Curriculum Development

While designing the KIBO curricula presented some challenges, the team of teachers and researchers were able to successfully implement three unique KIBO curricula in three different classrooms. The teachers themselves behaved as engineers by following the different stages involved in the engineering design process. They asked questions such as: What topic do I want to integrate with KIBO, what do I want my students to learn, and how can I integrate this project with the visual arts? They then imagined what their students might create, and planned out their curricular ideas, collaborating with the specialist teachers and researchers. Then, they tested out their plans as students worked on their final projects, revising and improving the plan as needed. Finally, teachers shared with one another about how the final projects went, and began the cycle again by asking questions about what could be done differently in the future.

Each teacher designed their class' final projects in a way to meet the unique needs of their students. First, each project focused on a curricular topic specific to the classroom. Teachers were given an opportunity to reflect on their current lesson plans and consider which one might be enhanced through the use of a new technology. As a pilot project, teachers did not have previous exposure to KIBO, so they could use this opportunity to explore one topic that they were already familiar with and test out what did and did not work. Additionally, this gave teachers an additional way to reinforce fundamental early education topics in a creative way using a new medium: robotics.

Second, the teachers adjusted the difficulty of the projects' goals based on the grade level of the students. For example, the second-grade teacher gave students the opportunity to create up to three projects, one for each of the worm environments. Additionally, she specifically designed the rock environment so that students would need to use a distance sensor, which is one of the more complex concepts of KIBO for children to understand and program. In contrast, the kindergarten project was much more straightforward. Children were asked to get their robot to travel in a straight line from Point A (one picture) to Point B (another picture). They were encouraged to experiment with repeat loops, but they could also successfully complete the program using basic programming blocks. Then, only once groups demonstrated that the robot traveled to Point B could they add extra instruction blocks. By adjusting the difficulty, students were able to successfully create personally meaningful projects.

The teachers discovered that creating appropriate curricula within the allocated time was not always as straightforward as expected. They learned how important it is to embrace not always knowing the "right answer," as well as to iteratively problem-solving along with their students. Over the course of the curricula, they had to adjust their initial approaches to the final project based on observations of their students and their past teaching experience. They needed to choose a focus for the final project that could translate well into the physical capabilities of KIBO, as well as keep their students engaged. Not all teachers use their initial curriculum idea. For example, the kindergarten teacher originally chose a different story instead of *Brown Bear, Brown Bear* for the basis of her curriculum. However, after rereading the story, she realized that the lack of plot would make it challenging for children to make creative and personally meaningful projects. She therefore had to go back to the drawing board and select a new book, which ended up integrating nicely with the robotics component. This teacher learned the value of changing an idea when it does not quite fit with the capabilities of the technology. She learned that technology has the power to bring literacy to life in a new way that is exciting for students. For the first-grade teacher, she noticed that having each group create four programs, each one representing a portion of the life cycle, would be too time-consuming and challenging, so she adjusted the goal to have students demonstrate mastery of one part of one cycle. This teacher learned about the unexpected time constraints that come with using complex technologies like robotics and how to adapt an initial curriculum idea to fit within the technological constraints.

11.8 Students' Learning

Children embraced the engineering design process as they went through the curricula. They had to plan out their programs in their Engineering Design Journals, test each program iteration, revise it multiple times to make it better, and then share it with others. They also engineered creative solutions to each of the challenges. The second-grade students could not have KIBO travel through a rock environment as it would have damaged the robots, but they were able to get the same point across by representing it with one big rock. For the first-grade classroom, students could not actually make their butterflies fly or their frogs hop, so they had to find other ways to represent these actions with KIBO's programming blocks. In the kindergarten classroom, students had to imagine what movements their book characters might do since it was not specified in the story.

Focusing specifically on programming, students mastered the syntax and rules of KIBO's programming language. Each grade experimented with sensors and advanced programming concepts in order to create more interactive and engaging projects. For example, kindergarten students created programs using repeat loops to minimize the number of Forward blocks that they would need for their programs. Additionally, they learned multiple ways to assemble sturdy and functional robots using motors, wheels, sensors, and lights. Their mastery of robotics was demonstrated at the end of the unit when each group had a functional KIBO robot and a syntactically correct program to share. During the presentations, each group was able to articulate their reasoning behind their programming and construction choices, as well as demonstrate their robots in action.

Not only were the students exposed to fundamental programming and engineering concepts and the intended final project topic, but they also engaged with a variety of other traditional and non-traditional school subjects. Every class incorporated the visual arts through their robot decorations created using craft and recyclable materials. Also, all groups had to use estimation, measurement, and counting to calculate the distance and direction they needed KIBO to move when creating their programs. Beyond traditional early childhood classroom subjects, students practiced collaboration by working in small groups, developed their presentation skills through sharing their work, and exercised perseverance in the face of challenging activities. As one teacher stated, the children were "all learners engaged in the project. There weren't outliers anywhere because they all found something that they were good at within the project and it make them feel really confident."

11.9 Conclusion

The engineering design process was a powerful concept that guided student and teacher learning through the curricula. While each teacher took a different approach to the final KIBO projects, all were generally successful at introducing the engineering design process, robotics, and programming with the support of Tufts and Lesley University researchers. It was helpful to allow teachers to draw on subjects they were already comfortable teaching (i.e., the natural sciences or literacy) as a bridge to implementing robotics for the first time. These vignettes illustrate how easily KIBO integrates with a variety of early childhood curricula and skills that children are naturally learning at that age. Additionally, it shows how KIBO can incorporate multiple subjects at once. For example, children may explore mathematical concepts such as estimation to program their robots while they engage in dramatic play imagining their robot acting out a famous story.

It is easy to see indicators of the children's success with KIBO in these vignettes—they were able to successfully program their robots and present complex work by the end of the curriculum. It is perhaps more difficult to see the learning process that the *teachers* engaged with throughout this experience. Much like the students immersed themselves in the engineering design process, the teachers also engaged with this iterative process of learning and experiencing failures before achieving success. Not only were they new to KIBO and faced with the challenge of mastering a new technology, but they were also new to designing integrative curricula tying in technology, engineering, and traditional early childhood content. These vignettes show the benefit of scaffolding the teachers' learning experience when embarking on a new technology initiative. In this case, this was done through support from the Tufts and Lesley research team, but it might also take the form of professional developments or collaborating with a school's technology specialist.

Looking forward, this school is now equipped with three early childhood teachers who are ready and capable of continuing engineering education in the early grades on their own and supporting new teachers joining this initiative. One teacher commented, "I was so excited about it [KIBO] I decided to share it with my colleagues. I gave them the opportunity to be the learners and had a little professional development with them so that they would feel comfortable and be able to overcome that barrier of being afraid to use it to being excited to use it. By the end of it, they were all saying they were really excited and can't wait to give it a try next year." Whether or not these young children grow up to be engineers, they have gained the problem-solving, collaboration, and perseverance skills necessary to excel in literacy, science, the arts, or any other area they may pursue in the future.

References

American Academy of Pediatrics. (2003). Prevention of pediatric overweight and obesity: Policy statement. *Pediatrics, 112,* 424–430.

Bers, M. (2008). *Blocks to robots: Learning with technology in the early childhood classroom*. New York: Teachers College Press.

Bers, M. U. (2010). The TangibleK robotics program: Applied computational thinking for young children. *Early Childhood Research and Practice, 12*(2).

Bers, M. U. (2012). *Designing digital experiences for positive youth development: From playpens to playground*. Oxford: Cary, NC.

Bers, M. U. (2018). *Coding as a playground: Programming and computational thinking in the early childhood classroom*. New York, NY: Routledge Press.

Bers, M. U., Flannery, L. P., Kazakoff, E. R., & Sullivan, A. (2014). Computational thinking and tinkering: Exploration of an early childhood robotics curriculum. *Computers and Education, 72,* 145–157.

Bers, M. U., Seddighin, S., & Sullivan, A. (2013). Ready for robotics: Bringing together the T and E of STEM in early childhood teacher education. *Journal of Technology and Teacher Education, 21*(3), 355–377.

Brophy, S., Klein, S., Portsmore, M., & Rogers, C. (2008). Advancing engineering education in P-12 classrooms. *Journal of Engineering Education, 97*(3), 369.

Brosterman, N. (1997). *Inventing kindergarten*. New York, NY: H.N. Abrams.

Clements, D. H. (1999a). The future of educational computing research: The case of computer programming. *Information Technology in Childhood Education Annual,* 147–179.

Clements, D. H. (1999b). Young children and technology. In *Dialogue on early childhood science, mathematics, and technology education*. Washington, DC: American Association for the Advancement of Science.

Druin, A., & Hendler, J. A. (2000). *Robots for kids: Exploring new technologies for learning*. Morgan Kaufmann.

Dweck, C. S. (2006). *Mindset: The new psychology of success*. New York: Random House.

Elkin, M., Sullivan, A., & Bers, M. U. (2014). Implementing a robotics curriculum in an early childhood Montessori classroom. *Journal of Information Technology Education: Innovations in Practice, 13,* 153–169.

Fessakis, G., Gouli, E., & Mavroudi, E. (2013). Problem solving by 5–6 years old kindergarten children in a computer programming environment: A case study. *Computers and Education, 63,* 87–97.

Kazakoff, E. R. & Bers, M. U. (2014). Put your robot in, Put your robot out: Sequencing through programming robots in early childhood. *Journal of Educational Computing Research, 50*(4).

Kazakoff, E., Sullivan, A., & Bers, M. U. (2013). The effect of a classroom-based intensive robotics and programming workshop on sequencing ability in early childhood. *Early Childhood Education Journal, 41*(4), 245–255. https://doi.org/10.1007/s10643-012-0554-5.

Lee, K., Sullivan, A., & Bers, M. U. (2013). Collaboration by design: Using robotics to foster social interaction in Kindergarten. *Computers in the Schools, 30*(3), 271–281.

Massachusetts Department of Education. (2006). Massachusetts science and technology/engineering curriculum framework. Retrieved from http://www.doe.mass.edu/frameworks/scitech/1006.pdf.

Massachusetts Department of Elementary and Education (2015). *Enrollment data.* Retrieved from: http://profiles.doe.mass.edu/profiles/student.aspx?orgcode=00490040&orgtypecode=6&.

Peel, S., & Prinsloo, F. (2001). Neuro-integration movements workshop: Module 1. *Brain Gym Edu-K South Africa.*

Resnick, M. (2003). Playful learning and creative societies. *Education Update, 8*(6). Retrieved from http://web.media.mit.edu/~mres/papers/education-update.pdf.

Resnick, M., Martin, F., Berg, R., Borovoy, R., Colella, V., Kramer, K., & Silverman, B. (1998). Digital manipulatives. In *Proceedings of the CHI '98 Conference.* Los Angeles, CA, USA.

Rogers, C. B., Wendell, K., & Foster, J. (2010). A review of the NAE report, engineering in K-12 education. *Journal of Engineering Education, 99*(2), 179–181.

Sesame Workshop. (2009). Sesame workshop and the PNC foundation joint White House effort on STEM education. Retrieved from http://www.sesameworkshop.org/newsandevents/pressreleases/stemeducation_11212009.

Strawhacker, A. L., & Bers, M. U. (2014). "I want my robot to look for food": Comparing children's programming comprehension using tangible, graphical, and hybrid user interfaces. *International Journal of Technology and Design Education.* https://doi.org/10.1007/s10798-014-9287-7.

Sullivan, A., & Bers, M. U. (2015). Robotics in the early childhood classroom: Learning outcomes from an 8-week robotics curriculum in pre-kindergarten through second grade. *International Journal of Technology and Design Education*, 1–18.

Sullivan, A., Elkin, M., & Bers, M. U. (2015). KIBO robot demo: Engaging young children in programming and engineering. In Proceedings of the 14th International Conference on Interaction Design and Children (IDC '15). ACM, Boston, MA, USA.

Svensson, A. (2000). Computers in school: Socially isolating or a tool to promote collaboration? *Journal of Educational Computing Research, 22*(4), 437–453.

White House. (2011). *Educate to innovate.* Retrieved from: http://www.whitehouse.gov/issues/education/educate-innovate.

Zelazo, P. D., Carter, A., Reznick, J. S., & Frye, D. (1997). Early development of executive function: A problem-solving framework. *Review of General Psychology, 1*(2), 198.

Chapter 12
Seeds of STEM: The Development of a Problem-Based STEM Curriculum for Early Childhood Classrooms

Mia Dubosarsky, Melissa Sue John, Florencia Anggoro, Susmitha Wunnava and Ugur Celik

Abstract This chapter adds to the body of research on engineering in early childhood education by describing the multiple research components associated with the development of an early childhood engineering curriculum, *Seeds of STEM*. Since very few research studies were devoted to the topic of engineering in early childhood, the *Seeds of STEM* research team was charged with developing many of the tools and instruments to be used throughout the project. The chapter describes the research conducted by the *Seeds of STEM* team in order to establish the framework for the curriculum, the development process, evaluation of fidelity of implementation, as well as the effectiveness of the curriculum. More specifically, the chapter addresses the following questions on curriculum development research: (a) Who should be part of the curriculum development team? (b) What is a successful curriculum development process? (c) What principles should guide the *Seeds of STEM* units? (d) How should the curriculum's effectiveness be measured? and (e) What measures should be taken to ensure fidelity of implementation?

12.1 Importance of Early Engineering

Improving science, technology, engineering, and mathematics (STEM) has become a great concern of researchers, educators, parents, and policy makers (Custer & Daugherty, 2009; Jenniches & Didion, 2009; Katehi, Pearson, & Feder, 2009; Schunn, 2009). The focus on STEM is largely related to issues concerning US competitiveness in the global economy and developing a skilled labor force with the ability to solve problem and address technological issues (CCNY, 2009; NSB, 2007). However, most efforts to improve STEM education in the USA have been

M. Dubosarsky (✉) · M. S. John · S. Wunnava · U. Celik
Worcester Polytechnic Institute (WPI), Worcester, MA, USA
e-mail: mdubosarsky@wpi.edu

F. Anggoro
College of the Holy Cross, Worcester, MA, USA
e-mail: fanggoro@holycross.edu

L. English and T. Moore (eds.), *Early Engineering Learning*, Early Mathematics Learning and Development, https://doi.org/10.1007/978-981-10-8621-2_12

limited to mathematics and science (and technology to a lesser degree)—all of which have a long history of established learning standards in the K-12 grades (Hsu, Purzer, & Cardella, 2011; National Research Council, 2011). Recently, engineering education has received increased attention at the K-12 levels (National Academy of Engineering, 2006; National Academy of Engineering & National Research Council, 2009; NGSS Lead States, 2013) and engineering education standards have been adopted by many states. There is also increased interest in the part of early childhood teachers and administrators in the importance of STEM education (Lindeman, Jabot, & Berkley, 2013; Moomaw, 2012; Wynn & Harris, 2013).

A child's academic success is largely dependent on a strong foundation for learning. The first five years of life are extremely significant for children's cognitive and skill development. During these years, young children explore their environments and use this information to develop language and construct abstract concepts and theories about the world around them (Bowman, Donovan, & Burns, 2001; Gelman, 1999; French, 2004; Worth, 2010). These early cognitive structures are the foundation for academic learning that includes further development of theories, strategies, and skills and are characterized as being deeply rooted in the child's environment and early interactions (Bowman et al. 2001; Eshach, 2006; French, 2004; Novak, 1977). The richness of the environment, type of interactions, and early experiences are linked to elaborate cognitive structures and better preparedness for further learning. In fact, high-quality preschool education has been found to significantly improve young children's learning outcomes (Camilli, Vargas, Ryan, & Barnett, 2010; Gorey, 2001; Gormley, Phillips, & Gayer, 2008). In addition, children constantly explore and question the mathematical and scientific world around them (Bowman et al., 2001; Brenneman, Stevenson-Boyd, & Frede, 2009; French, 2004; Gelman, 1999; Worth, 2010), making this period of early childhood ideal for introducing concepts in STEM and engaging children in developmentally appropriate activities to begin understanding the world around them.

Early *engineering* education, the E in STEM, involves the systematic process of problem solving, which is important for several reasons. First, children have been described as 'born engineers' (Cunningham, 2009). Engineers are problem solvers, and the engineering design process requires creativity, collaboration, and communication. Children reason, define problems, manipulate, build and test prototypes, apply mathematical and scientific concepts, and share their newfound solutions with friends and family (Christenson & James, 2015). Research shows that when teachers engage in the engineering design process, children increase in their engagement of activities, the number of engineering behaviors displayed, and their persistence in completing activities (Wang et al., 2013). By incorporating engineering concepts into early education, children are provided with developmentally appropriate knowledge and skills to further examine the world (Ackerman & Barnett, 2005; Brenneman et al., 2009).

Second, although engineering overlaps with science, technology, and math, which are mostly covered in a typical preschool curriculum (Bagiati & Evangelou, 2015), some skills and concepts are specific to the engineering field (Schunn, 2009). Engineering is a context-based subject with real-life applications, which appeals to a diverse group of students and therefore serves as an anchor for deepening scientific knowledge (NGSS Lead States, 2013). This notion fits well with the call put forth by early childhood and cultural education scholars to develop an early childhood science curriculum that is connected to children's lives and experiences. Context-based activities have been shown to engage and include all learners, thus leading to increased motivation and achievement (Bowman et al., 2001; Lynch, 2001; New, 1999; Warren, Ballenger, Ogonowski, Rosebery, & Hudicourt-Barnes, 2001). Integrating engineering into the early childhood curriculum gives children the opportunity to examine these everyday concepts and build upon them in a new way (Bagiati & Evangelou, 2015). Research on engineering design projects finds that students show improved problem-solving skills that are critical in dealing with ambiguity and solving open-ended and ill-defined problems (Eshach, 2006) and enhance students' content knowledge and skills in science (Kolodner et al., 2003; Mehalik, Doppelt, & Schunn, 2008) and mathematics (Hjalmarson, Diefes-Dux, & Moore, 2008). For preschoolers, such opportunities can increase persistence in completing activities (Cohen & Uhry, 2011) and better critical thinking and social skills (Stoll, Hamilton, Oxley, Eastman, & Brent, 2012; Tunks, 2009).

Third, stimulating interest and early exposure to engineering as a career requires the practice of engineering (Schunn, 2009). Studies show that engineering design projects enabled more positive attitudes toward engineering as a career (Cunningham & Lachapelle, 2010; Kolodner et al., 2003; Mehalik et al., 2008).

Finally, exposure to different problems and the application of science, math, and technology has the ability to reduce the achievement gap, while simultaneously debunk stereotypes and change attitudes and beliefs toward the STEM fields (Brophy, Klein, Portsmore, & Rogers, 2008). National reports highlight the wide gap in STEM literacy between low-income and ethnic minority American students and their White, middle-class American peers. This disparity is well documented by research from kindergarten to 12th grade, suggesting that gaps in academic and skill development start during the prekindergarten years. These achievement and readiness gaps (in reading, math, science, and approaches to learning) are evident as early as kindergarten and widen as students advance in school (Duncan et al., 2007; Clasessens, Duncan, & Engel, 2009; Federal Interagency Forum on Child and Family Statistics, 2013). Additionally, although girls and boys take roughly the same number of classes in elementary, middle, and high school and are equally prepared to pursue science and engineering majors in college (Hill, Corbett, & St. Rose, 2008; U.S. Department of Education, National Center for Education Statistics, 2007), fewer girls than boys decide to major, are retained, and actually go into these fields (Seymour & Hewitt, 1997; Xie & Shauman, 2003).

Stereotypes of ethnic minorities and women are partly responsible for these disparities (Eccles, Jacobs, & Harold, 1990; Fennema & Sherman, 1977; Jacobs & Eccles, 1985; Swim, 1994). Stereotyping is not uncommon for 3- to 6-year-olds (Levitch & Gable, 2016; Piaget, 1961), and this may lead to a fixed mind-set. Early childhood educators can help children combat stereotypes in the classroom through increased representation of engineers, engineering problems, and engineering occupations (Care, Denas, & Brown, 2007; Sleeter & Grant, 1999), through the language used and the literature selected (Roberts & Hill, 2003; Southern Poverty Law Center, 1997).

In summary, introducing STEM and especially engineering during early childhood education promises to provide young children with the problem-solving skills that will help them address complex problems and better prepare them for success in school and life. Early childhood teachers who teach children to solve problems in a systematic way can also introduce children to possible career opportunities while debunking stereotypes.

12.2 Research on Engineering Education During Early Childhood

Despite the promising evidence that introducing STEM/engineering ideas and practices during the early childhood years supports children's cognitive development and positive attitudes toward learning and inquiry (Eshach, 2006; Evangelou, 2010; Katz, 2010; Van Meeteren & Zan, 2010), there is very little STEM or engineering instruction within the prekindergarten classrooms (Diamond, Justice, Siegler, & Snyder, 2013; Ginsburg, Lee, & Boyd, 2008).

One of the reasons for the lack of STEM and engineering instruction is teachers' low self-efficacy regarding the teaching of STEM, due in part to a lack of preparation and shortage of early childhood STEM and engineering curricula (Bagiati, Yoon, Evangelou, & Ngambeki, 2010; Brenneman, 2011; Greenfiled et al., 2009; New, 1999).

Research on STEM and engineering in early childhood education is also limited. At the time of writing this chapter, only a few research projects engage in systematic study of STEM and engineering during the early childhood years, shedding light on teaching and learning as well as classroom materials and curricula. These studies include the development of the Engineering is Elementary (EiE) curriculum (Cunningham & Lachapelle, 2007, 2010) and the development of a pre-K engineering curriculum (e.g., Bagiati & Evangelou, 2015). Additional early childhood STEM/engineering research themes include robotics activities (Sullivan, Kazakoff, & Bers, 2013), STEM during summer camp (Torres-Crespo, Kraatz, & Pallansch, 2014), and a recent guide of PreK-5 engineering curricula (Sneider, 2014).

This chapter adds to the body of research on engineering in early childhood education by describing the multiple research components associated with the development of an early childhood engineering curriculum, *Seeds of STEM*. Since very few research studies were devoted to the topic of engineering in early childhood, the *Seeds of STEM* research team was charged with developing many of the tools and instruments to be used throughout the project. The following sections describe the research conducted by the *Seeds of STEM* team in order to establish the framework for the curriculum, the development process, evaluation of fidelity of implementation, as well as the effectiveness of the curriculum. More specifically, the chapter addresses the following questions on curriculum development research:

1. Who should be part of the curriculum development team?
2. What is a successful curriculum development process?
3. What principles should guide the *Seeds of STEM* units?
4. What measures should be taken to ensure fidelity of implementation?

12.3 About *Seeds of STEM*

Teacher (showing a puppet): Good morning, friends! Today we have a special visitor. Its name is Problem Panda, and it needs your help. Do you want to meet our guest?
Children: Yes!
Puppet (in a sad voice): Hello, children! My name is Problem Panda and I have a big problem. I heard that you are learning to solve problems just like engineers, and I thought perhaps you could help solve my big problem. Can you help me?
Children: Yes, we can help you.
Puppet (happy): Oh, thank you! Yesterday, I was getting ready to go and meet my friend and I bought a special present for her—a ring. Somehow the ring fell into a glass of water and was frozen—see? (showing a cup with a ring stuck inside an ice chunk). Now I can't take the ring out—what will I do?
Teacher (to children): Let's put on our 'engineer' badges and help Problem Panda.

The vignette above is taken from a video of a Head Start classroom (3- to 5-year-old children) in Worcester, Massachusetts, testing the second unit of the *Seeds of STEM* curriculum. *Seeds of STEM* is an innovative, research-based curriculum which focuses on teaching preschool children—and their teachers—the process of problem solving. The different units of the curriculum are built around the steps of the engineering design process (EDP). An accompanying character, Problem Panda, presents to the children a different problem in every unit and engages the children in the process of understanding the problem and defining criteria for successful solutions, brainstorming solutions, selecting and testing some of the solutions, creating and revising the solution, and sharing the solution with Problem Panda and other guests. In addition to teaching children how to solve problems, each unit addresses a key concept (core idea) in science. For example, the science focus of unit 2, from which the vignette is taken, is **ice and water, solids and liquids**, and the engineering focus is **brainstorming and selecting testable solutions**. During the first week of the unit, the children engage in multiple experiences about the ice and water, solids and liquids, including stories, melting and freezing experiments, sorting, going on a solid/liquid hunt, and even freeze-dancing. During the second week, Problem Panda asks the children to help him with a problem: get a valuable item (the ring) out of the ice without harming it in the process. With guidance from the teacher, the children practice brainstorming solutions to Panda's problem. The teacher encourages the children to propose different solutions based on children's experimentation with ice

Table 12.1 Seeds of STEM unit description

Unit	Science week	Engineering week	Main problem
1	Introduction to the problem-solving process		Help Panda get out of a box
2	Ice and water (solids and liquids)	Identify problem, brainstorm, sort, and vote on solutions	Panda dropped a ring into a cup of water that froze! Help Panda get the ring out of the ice
3	Habitats	Plan and create models	Panda's friend is coming to visit! Plan a habitat for Sally Squirrel
4	The five senses	Test and improve solutions	Panda wants to play with his friend Design a toy for a blind friend
5	Forces and motion	Share solutions with others	Panda broke his leg! Design a device that helps Panda move
6	Properties of materials	The entire process	Design a container to send cookies to a friend who lives across the river
7	Plant parts and needs	The entire process	Gladys Goat ate Panda's plant! Design a barrier to protect plants
8	Light and shadow	The entire process	Panda wants to play outside, but it is too hot and bright! Design a shade for Panda

and water during the first week, emphasizing that every problem has multiple solutions. The children then define criteria for successful solutions (e.g., not breaking the ring, melting the ice fast) and sort the solutions into testable and non-testable in the classrooms. The children then vote on a solution they would like to test first.

Table 12.1 presents the science and engineering focus for each Seeds of STEM unit.

The *Seeds of STEM* development process offers a unique model of collaboration between Worcester Head Start (WHS) teachers and the research team, representing researchers from Worcester Polytechnic Institute (WPI) and College of the Holy Cross. During the first two years of the project, the curriculum was developed through an iterative process of creation, testing, and revision. Currently, during the third year of the study, the curriculum is being pilot-tested at a different Head Start program. The project is supported by a grant from the US Department of Education's Institute of Education Sciences (IES, grant # R305A150571) and expected to be completed in 2019.

The overarching goal of the project is to support the teaching and learning of STEM practices in early childhood and, as a result, increase students' STEM readiness. The curriculum is developed to achieve two main student learning outcomes:

(1) Children who experience the *Seeds of STEM* curriculum will demonstrate improved ability to appropriately use STEM vocabulary that is integral to the engineering design process.
(2) Children who experience the *Seeds of STEM* curriculum will demonstrate improved ability to conduct each step of the engineering design process, which includes the following: define/explain a problem in their own words, propose multiple solutions to solve the problem, test and improve one solution of choice, and communicate the solution others.

1. **Who Should Be Part of the Curriculum Development Team?**

Education research relies on partnerships between researchers and practitioners to inform practice and enrich the research (Clements, 2007; NRC, 2002). Therefore, a robust collaboration between the research team and the Worcester Head Start program was established.

The *Seeds of STEM* curriculum development team included experts in STEM education, engineering, cognitive development, and diversity, as well as lead teachers from the Worcester Head Start program who provided classroom and pedagogy expertise. The development process was overseen by an advisory board consisting of early childhood, social science, and engineering experts. The team met monthly during the curriculum development period.

We believe that this collaborative development model is crucial for the success of the curriculum. The teachers on the team are knowledgeable about classroom environment, pedagogy, individual children's abilities and interests, appropriate vocabulary, attention span, and family involvement. They are also familiar with the daily schedule, the available materials, and other requirements of Head Start programs around the country. Teachers' expertise, combined with the research team's knowledge in STEM and children's learning, ensured the accurate matching of the curriculum's activities to a real-classroom environment, enhancing feasibility of implementation and curriculum usability by teachers.

To select teachers for the development team, the researchers defined the following selection criteria: (1) having a master's degree, (2) working in Worcester Head Start for more than one year, (3) mid to high scores on all Classroom Assessment Scoring System (CLASS) dimensions, (4) good reviews from the teacher's supervisor, and (5) teachers' interest in joining the development team. Based on these criteria, seven lead teachers from Worcester Head Start were selected for the role of 'Developer Teachers' (DT). The relationship among team members was one of mutual respect. During unit development meetings, the researchers presented the expected outcomes for students and teachers, while the teachers proposed activities, stories, and tasks that lead to such outcomes. Two of the developer teachers also participated in the final revision of the curriculum, to help make it as 'teacher-friendly' as possible.

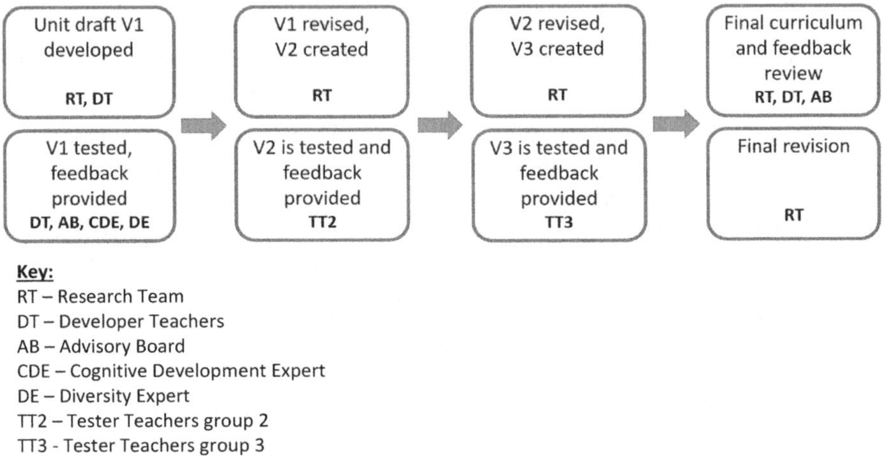

Fig. 12.1 Seeds of STEM curriculum development process

Lastly, the collaboration provided professional growth opportunity for the developer teachers. Several of the teachers presented at state and national conferences and participated in the facilitation of *Seeds of STEM* professional development sessions.

2. **What Is a Successful Curriculum Development Process?**

Research on curriculum development has shown that an iterative approach to development, in which each unit is developed, tested, revised, and tested again, leads to a high-quality curriculum (Clements, 2007; Diamond & Powell, 2011; Kinzie, Pianta, Kilday, McGuire, & Pinkham, 2009). This approach is in line with the engineering design process (EDP) that calls for engaging in a systematic and repetitive process of testing and revision until the final solution is ready for use.

In accordance with the iterative approach, each unit of the *Seeds of STEM* curriculum was tested and revised three times during the development process by three groups of Head Start teachers: the **developer teachers (DT)** and two groups of **tester teachers (TT2 and TT3)**. Figure 12.1 provides a visual of the entire process.

The first draft of each unit (V1) was developed collaboratively with the developer teachers, with the research team defining the standards and learning outcomes to be addressed, and the teachers proposing books, tasks, activities, songs, and art projects to meet the defined outcomes. Once the first draft was created, each one of the DT tested the unit in their classrooms and provided detailed feedback about the activities, the props, and the engagement of their children with each activity in the unit. During the testing of the first draft, the advisory board met to provide feedback on the unit; feedback was also provided by the diversity and cognitive development experts on the team.

The research team analyzed the feedback from the DT and experts and created the second version of the unit (V2). The group of **tester teachers (TT2)** tested the revised version of each unit in their classrooms and provided detailed feedback similar to the DT. The feedback from TT2 was analyzed and compared with video data from the classrooms and used for the second revision of each unit. Once all units have been tested and revised twice, the development team reviewed all the units together to ensure a cohesive flow from one unit to the next. This step proved to be very important. For example, the character of Problem Panda was only created during the development of unit 3 (earlier versions of unit 2 had a character named 'Mr. Problemo'). Once all units were developed, it became clear that Problem Panda should be present in units 1 and 2 as well. Revisions were made, and the third version of the curriculum (V3) included Problem Panda as a leading character from the first unit.

Once revised, a third group of **tester teachers (TT3)** tested the entire curriculum in sequence (from beginning to end), in order to assess the cumulative learning outcomes for children and the flow of the entire curriculum. The group of TT3 was asked to provide detailed feedback about the clarity of instructions and student outcomes.

Following the third testing and feedback from TT3, the development team and advisory board finalized the curriculum (V4), to be tested in experimental study by teachers from a different Head Start program.

3. **What Principles Should Guide the Development of *Seeds of STEM* Units?**

Following an extensive review of the literature, the *Seeds of STEM* research team defined a set of eight research-based principles to guide the development of the curriculum and ensure its high quality. The team adapted the Dayton Regional STEM Center's Quality STEM Framework (2011) to meet the standards for high-quality early childhood education. A description of the each of the guidelines follows Table 12.2.

Developmentally appropriate. Head Start classrooms are comprised of children of various ages and often include a wide range of skills, abilities, and language backgrounds (Cabell, Justice, Konold, & McGinty, 2011). To ensure that each unit of the curriculum is developmentally appropriate, the team relied on the National Head Start Child Development and Early Learning Framework (2010), and the Massachusetts Framework for Science, Technology, and Engineering for Pre-K (2014) to define the learning outcomes of the curriculum. The developer teachers proposed activities and tasks that cater to their multi-age and multi-ability classrooms. Through the iterative process of trial, feedback, and revisions, we were able to select only the activities that were proven to engage all children, including children who are dual language learners (DLLs). To increase engagement with the curriculum, the development team created a character, Problem Panda (exemplified by a stuffed animal), that visits the children in each unit to present a new problem.

Culturally responsive. Research shows that cultural contexts affect young children's cognitive, social, and emotional development, as well as their approaches to learning (Bowman et al., 2001; Genishi & Goodwin, 2008). A school's culture may differ greatly from a minority group's home culture. New (1999) called for early childhood teachers to embrace children's home culture and model the coexistence

Table 12.2 Guidelines for high-quality STEM experiences for early childhood classrooms

1. Developmentally appropriate	Seeds of STEM learning experiences provide children with books, videos, materials, and tasks that are appropriate for their cognitive and language development
2. Culturally responsive	Seeds of STEM learning experiences are designed to reflect diversity of gender, ethnic background, and physical abilities while allowing access and emphasizing children's own culture
3. Application of the engineering design process	Seeds of STEM learning experiences engage children in an **open-ended, multiple solutions problem-solving task** which requires them to follow the **engineering design process** (i.e., defining the problem, brainstorming, researching, creating, testing, improving, and communicating)
4. Integrity of the academic content	Seeds of STEM learning experiences are content-accurate, aligned with the relevant **content standards,** and **foundational skills** of **Science, Technology**, **Engineering**, and **Math** as articulated in pre-K standards and frameworks
5. Quality of technology integration	Seeds of STEM learning experiences are **hands-on** in nature and require children to use variety of **tools** to solve each problem (e.g., scissors, scales, computers, rulers, hand lenses)
6. Connections to Non-STEM disciplines	Seeds of STEM learning experiences help children connect STEM knowledge and skills with standards from early literacy, art, social, emotional, and physical education
7. Real-world connections and STEM careers	Seeds of STEM learning are driven by a real-world phenomena, which are familiar and relevant to the children's life inside and out of the classroom When applicable, quality STEM learning experiences introduce different STEM careers and help children understand the roles of people who work in STEM disciplines
8. Nature of assessment	Seeds of STEM units include formative and summative authentic embedded assessments. The variety of activities allow children to demonstrate their understanding in different ways and allow teachers to record children's mastery of learning outcomes

of different cultures. This allows children of minority cultures to value their home culture and the school (majority) culture and learn to celebrate the differences in people's identities. Cultural-based education recognizes the language, experiences, values, and knowledge of children, their families, and their communities and includes elements of children's home culture into the daily curriculum (Dubosarsky et al., 2011). The *Seeds of STEM* curriculum addresses cultural responsiveness by using books, images, and scenarios that represent a diversity of cultures, allowing children to identify and feel included in the units, while learning to respect other cultures. In addition, research suggests that parent–teacher collaboration supports student learning, and parental involvement is associated with academic and social competence (Powell, Son, File, & San Juan, 2010). To build on these findings, each unit plan includes extension activities and home-connection ideas for engaging the family with the topic of the unit.

Application of the Engineering Design Process. The research team strongly believes that the process of problem solving is the heart of STEM education, and therefore, the process should be taught explicitly through the curriculum. The problems that are presented to the children should be open ended and allow for multiple solutions. To support the curriculum's mission of teaching children how to solve problems, and with full understanding that the process may be too abstract for young children, *Seeds of STEM* engages the teachers in creating a visual aid of the problem-solving process. The visual, which looks differently in each classroom, is introduced during the first unit and being referred to multiple times in every unit of the curriculum. The teachers created the problem-solving visuals during a workshop, using vocabulary words and images that fit their students' understanding. It was important to the research team that every classroom team will create their own visual, thus increasing the ownership of teaching the process. A few examples of teachers' visuals are found in Fig. 12.2.

Integrity of academic content. Based on the authors' experience of conducting STEM professional development workshops with P-12 teachers, often times STEM projects are designed as 'add-on' experiences, for example 'egg drop' or bridge-building challenges, without making clear connection between the STEM challenge and the instructional core ideas (academic standards). The *Seeds of STEM* problems and tasks are aligned with science, math, or literacy standards, in order to make connections between STEM (engineering in particular) and the rest of the pre-K curriculum and prepare children to become problem solvers in any subject. As described earlier, each one of the *Seeds of STEM* units is aligned with NGSS core ideas, as well as the Massachusetts Framework for Science, Technology, and Engineering for Pre-K, and the Head Start Framework. The problems presented to the children require them to apply the science and math concepts for creating successful solutions.

Quality of technology integration. A common misconception held by educators (and the general public) narrows the definition of technology to digital technology. However, the technology integration principle calls for using tools, any kind of tool, that children find useful in solving the problem. These tools may include scissors, a scale, child-sized hammer, marker, measuring tape, spoon, camera, as well as a computer or tablet for research. Teachers who follow the *Seeds of STEM* curriculum

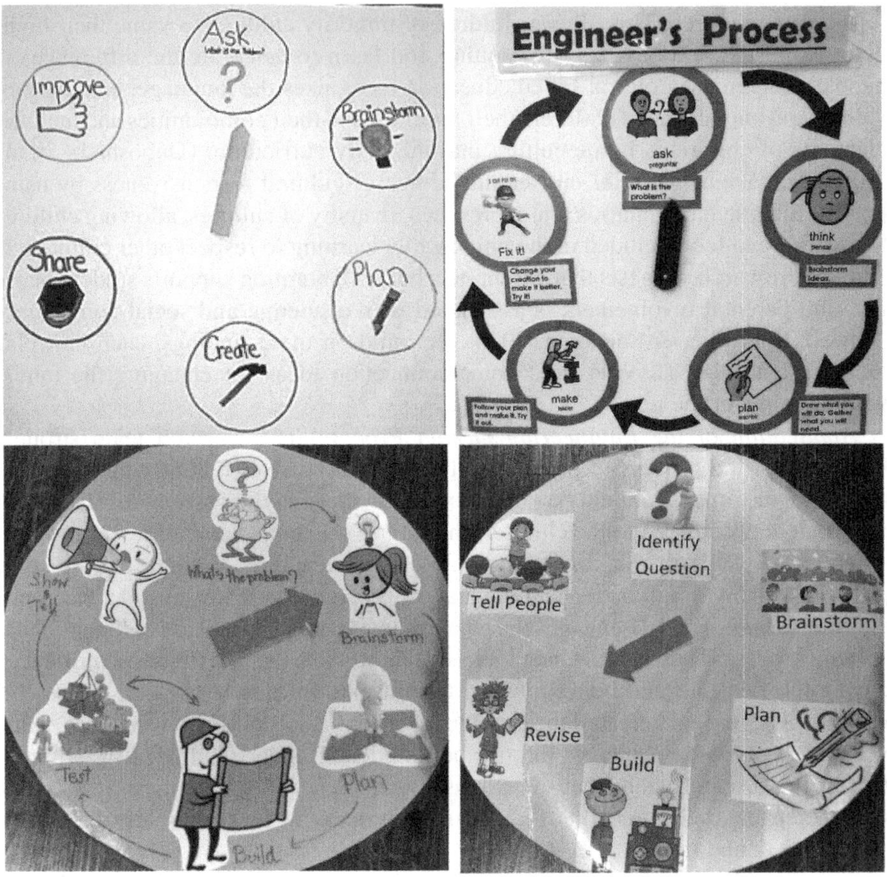

Fig. 12.2 Teacher-created visuals of the problem-solving process

consider the tools that children can use safely and include a plan on teaching the children how to use these tools.

Connection to non-STEM disciplines. Integration is a key to STEM education and the curriculum reflects that by using non-STEM context, such as books, videos, stories, and social studies topics, to generate problems. A study involving kindergarten students found that integration of science and literacy increased children's motivation and engagement in science (Samarapungavan, Mantzicopoulos, & Patrick, 2008). William Wolfson's *Engineering Lens* method (http://www.integratingengineering.org/index.html) for integrating engineering practices with literacy was used by the research team during the development process. There are several advantages for the integration of STEM with literacy: first, greater engagement from teachers, who are very comfortable with literacy activities. Second, this approach may help overcome common stereotypes (i.e., engineering is building). Third, using books as a context to start the problem-solving process allows for the choice of books that represent real-world relevance and involving a diverse population of characters. Fourth, adding books as the context for problem solving assists with daily lesson planning since it is expected to address literacy daily.

In addition to integration of STEM with literacy, the *Seeds of STEM* curriculum includes arts, music, and physical activities. The children go on a force hunt (forces and motion), shadow hunt (light and shadow), and solid/liquid hunt (ice and water) around the school and end many of the daily activities with 'freeze dance' that helps emphasize some of the concepts through kinesthetic learning.

Authentic assessment. Since the *Seeds of STEM* curriculum is centered on STEM practices, authentic assessments that are embedded in the unit would measure children's mastery of learning outcomes. Each unit plan includes formative assessment tasks that ask children to explain, demonstrate, or design a solution to a problem. The teachers document children's learning by scribing their explanations on the plans. The teachers also record children's demonstration of the problem-solving steps and use of vocabulary in context. Since young children may show evidence of transfer of the learning during other parts of the day, each unit includes a checklist of learning outcomes and the teacher is able to record evidence for learning—such as using the unit's vocabulary or demonstrating a key skill—throughout the day.

The repetitive nature of the curriculum, and the emphasis of the problem-solving process, engages the children in solving problems with different contexts. The last unit of the curriculum also serves as authentic summative assessment, allowing the teacher to evaluate children's mastery of the curriculum's learning outcomes.

4. **What Measures Should Be Taken to Ensure Fidelity of Implementation?**

Following the completion of the development process, the *Seeds of STEM* curriculum is being tested in 17 Head Start classrooms to assess its effectiveness in meeting the defined student outcomes. In order to evaluate the extent to which the curriculum was implemented as intended, the team developed measures to gauge the fidelity of curriculum implementation (FoI). The team defined the problem-solving steps and the way these steps are being introduced and followed as the critical components of the curriculum, and developed two methods to evaluate *Seeds of STEM*'s FoI: teacher surveys and observation form.

(1) Teacher survey. During the pilot study, intervention teachers complete a survey for each unit they teach. In the survey, the teachers provide feedback on the activities they taught and describe modifications they made during implementation, including changes to pedagogical strategies or in materials used. Teachers also report the level of engagement that their students present in each unit's critical component.
(2) FoI Observation form. Each *Seeds of STEM* unit is being videotaped by the classroom teachers. The video recordings are analyzed using the FoI observation form. The form development process was initiated by aligning the critical components with the following areas of curriculum implementation: materials and resources, duration of activity, format of activity, conceptual accuracy, use of vocabulary words, teacher's efforts to engage children, student participation,

and teacher's efforts to promote inclusion. These areas were then compared to the FoI elements found in the literature: adherence, duration and exposure, quality of delivery, program specificity, and student responsiveness, as well as with the categorization into structural and instructional components (Century, Rudnick & Freeman, 2010; O'Donnell, 2008). Through an iterative process, the areas were narrowed, and indicators for each area of implementation were refined. An initial seven-point scale was later revised into four categories: no evidence, low fidelity, medium fidelity, and high fidelity. An iterative process of developing detailed range descriptors for each area of implementation/indicators is now complete. In developing the range descriptors, the team worked through examples, examined sample videos to see if the range descriptors captured the information accurately, and modified range descriptions accordingly until an agreement was reached. The goal of this process was to create FoI descriptors that include concrete and observable actions, behaviors, and language.

5. **How Do We Measure the Curriculum's Effectiveness?**

The *Seeds of STEM* curriculum will be considered 'effective' if children who participate in the pilot study demonstrate the following goals: (1) improved ability to use the problem-solving vocabulary in context and (2) improved ability to conduct each step of the engineering design process.

The first step in developing the assessment was to clearly define the learning outcomes for each unit, as well as for the entire curriculum. To do so, the research team reviewed three sets of education standards: The Next Generation Science Standards (NGSS), the Massachusetts Framework for Science, Technology, and Engineering for Pre-K, and the Head Start Framework. Analysis of these frameworks resulted in a defined list of problem solving practices for young children, which became the curriculum's learning outcomes. In addition, a list of key problem-solving vocabulary words was defined and embedded in each unit (Table 12.3).

Assessment of curriculum effectiveness (in the form of mastery of learning outcomes by the children) is conducted formatively, during each unit of the curriculum, and summatively, during unit 8.

Formative assessment methods. Two main methods of formative assessment are being employed as part of the *Seeds of STEM* curriculum: (1) a teacher checklist and (2) a rubric for coding observational data. The teacher checklist includes the specific outcomes and vocabulary words associated with each unit. Pilot teachers are asked to keep the checklist handy in the classroom and check if they notice that one or more children correctly use a vocabulary word or show evidence of mastering a step in the engineering design process. The checklist includes a space for comments, and teachers are able to provide the context and evidence for their marking. The rubric for observational data is currently in development. The rubric includes detailed descriptors that provide observational evidence for learning outcome mastery, in the form of expressive language, gestures, and student work. Specific activities within each unit include formative assessment questions and tasks, to be evaluated by the research team using the rubric.

Table 12.3 Seeds of STEM learning outcomes

	Outcome	Definition	Assessment	Outcome addressed in unit
1	Identify a problem	• Articulate a problem and its implications	• Formative • Summative	1–8
2	Ask questions	• Explore issues related to the phenomenon and problem	• Formative • Summative	1–8
3	Obtain information	• Use first-hand interaction with objects and organisms, media, and books to gather information • Collect and document information (using senses and tools, including technology, to gather information)	• Formative • Summative • Artifacts	2–8
4	Analyze information	• Discuss the meaning and value of information for solving problems • Articulate processes and relationships	• Formative • Summative	2–8
5	Brainstorm and propose solutions	• Draw on self and others' knowledge and observations to come up with multiple solutions	• Formative • Summative • Artifacts	2–8
6	Choose a feasible solution	• Review and organize the ideas • Classify potential solutions into: ordinary, innovative, and magical • Predict outcomes and anticipate difficulties	• Formative • Summative	2–8
7	Plan to execute solution	• Develop a plan for the design of the solution using simple materials/equipment • Investigate materials as needed	• Formative • Summative • Artifacts	3–8
8	Design/build a model of the solution	• Work with others to select and use materials to build the solution	• Formative • Summative • Artifacts	3–8
9	Test solution	• Implement the design • Gather data on the effectiveness of the solution	• Formative • Summative	4–8
10	Evaluate solution	• Assess the effectiveness of the solution in solving the problem • Identify limitations of the solution	• Formative • Summative	4–8

(continued)

Table 12.3 (continued)

	Outcome	Definition	Assessment	Outcome addressed in unit
11	Improve solution	• Address limitations by modifying existing solution or developing a new solution • Retest and evaluate modified solution • Repeat steps 9–11 as necessary	• Formative • Summative	5–8
12	Constructing explanations	• Look for and describe patterns and relationships between the solution, limitations, and the problem, focusing on cause-and-effect relationship • Draw conclusions based on evidence	• Formative • Summative	All
13	Share findings	• Communicate the entire problem-solving and design process • Articulate findings and conclusions to peers and teachers	• Formative • Summative	All
14	Vocabulary	Understand and use in the context of the following words: engineer, problem, solution, brainstorm, choose, plan, model, create, test, improve, share	• Formative • Summative	

Summative assessment methods. The last unit of the curriculum will serve as summative assessment. While the first five units introduce and guide children through the process of problem solving, units 6 and 7 allow the children to practice the entire problem-solving process, and finally unit 8 presents a novel problem and allows for authentic assessment of children's transfer of learning. Trained observers will use a cognitive coding rubric to assess seven dimensions based on the outcomes checklist (describe/recognize information, identify problem, obtain information/ask questions, brainstorming, solution planning, solution creating and testing, sharing findings). This rubric allows the coder to record both the frequency of a behavior, such as asking a clarifying question, and the level of sophistication within each dimension. The rubric also allows the coder to record how often target vocabulary words are repeated and used correctly in a novel context.

A second summative assessment was developed to evaluate children's recognition of vocabulary words and the steps of the problem-solving process. This individual assessment was developed as a short computer game which asks children to identify pictures of problems and solutions, sequence images of a problem-solving story, and identify a picture for each vocabulary word. Every child in the intervention and control classrooms will play the 'game' before and after experiencing the full *Seeds of STEM* curriculum.

This chapter described the extensive research associated with the development of the *Seeds of STEM* curriculum. The authors hope that the information about the development team, the iterative development process, the curriculum guidelines, evaluation of fidelity of implementation, and assessment of the curriculum effectiveness will add to the body of knowledge about STEM and engineering education research during the early childhood years.

The chapter was initially written during the curriculum development process and revised to include information about the pilot testing of the curriculum. Currently, the *Seeds of STEM* project is in its third year, and several of the research instruments are being finalized. For the most up-to-date information and publications, please visit the project Web site: www.seedsofstem.org.

Acknowledgements The research reported here was supported by the Institute of Education Sciences, US Department of Education, through Grant R305A150571 to Worcester Polytechnic Institute. The opinions expressed are those of the authors and do not represent views of the Institute or the US Department of Education.
The authors would like to thank Dr. Katherine Chen and Leah Reppucci for their thoughtful comments and feedback.

References

Ackerman, D. & Barnett, S. (2005). Prepared for kindergarten: What does "readiness" mean? *National Institute for Early Education Research Policy Brief*.

Bagiati, A., & Evangelou, D. (2015). Engineering curriculum in the preschool classroom: The teacher's experience. *European Early Childhood Education Research Journal, 23*(1), 112–128.

Bagiati, A., Yoon, S. Y., Evangelou, D., & Ngambeki, I. (2010). Engineering curricula in early education: Describing the landscape of open resources. *Early Childhood Research and Practice, 12*(2), 1–15.

Bowman, B. T., Donovan, M. S., & Burns, M. S. (Eds.). (2001). Eager to learn: Educating our preschoolers. Washington D.C.: National Academies Press.

Brenneman, K. (2011). Assessment for preschool science learning and learning environments. *Early Childhood Research and Practice 13*(1).

Brenneman, K., Stevenson-Boyd, J., & Frede, E. (2009). Math and science in preschool: Policies and practices. *National Institute for Early Education Research Policy Brief.*

Brophy, S., Klein, S., Portsmore, M., & Rogers, C. (2008). Advancing engineering education in P-12 classrooms. *The Research Journal for Engineering Education,* 369–387.

Cabell, S., Justice, L. M., Konold, T., & McGinty, A. (2011). Profiles of emergent literacy skills among preschool children who are at-risk for academic difficulties. *Early Childhood Research Quarterly, 26*(1), 1–14.

Camilli, G., Vargas, S., Ryan, S., & Barnett, W. S. (2010). Meta-analysis of the effects of early education interventions on cognitive and social development. *Teachers College Record, 112*(3), 579–620.

Care, E., Denas, J., & Brown, R. (2007). The realism and sex type of four- to five-year-old children's occupational aspirations. *Journal of Early Childhood Research, 5,* 155–168.

CCNY (Carnegie Corporation of New York). (2009). The opportunity equation—Transforming mathematics and science education for citizenship and the global econ-

omy. Available https://www.carnegie.org/media/filer_public/80/c8/80c8a7bc-c7ab-4f49-847d-1e2966f4dd97/ccny_report_2009_opportunityequation.pdf. Accessed February 18, 2016.
Century, J., Rudnick, M., & Freeman, C. (2010). A framework for measuring fidelity of implementation: A foundation for shared language and accumulation of knowledge. *American Journal of Evaluation, 31*(2), 199–218.
Christenson, L. A., & James, J. (2015). Building bridges to understanding in a preschool classroom: A morning in the block center. *Young Children*, 26–31.
Clasessens, A., Duncan, G., & Engel, M. (2009). Kindergarten skills and fifth grade achievement: Evidence from the ECLS-K. *Economics of Education Review, 28,* 415–427.
Clements, D. H. (2007). Curriculum research: Toward a framework for 'research-based curricula'. *Journal for Research in Mathematics Education, 38,* 35–70.
Cohen, L., & Uhry, J. (2011). Naming block structures: A multimodal approach. *Early Education Journal, 39,* 79–87.
Cunningham, C. (2009). Engineering is elementary. *The Bridge, 39*(3), 11–17.
Cunningham, C. M., & Lachapelle, C. P. (2007). *Engineering is Elementary: Children's changing understandings of science and engineering*. Paper presented at the American Society for Engineering Education Annual Conference & Exposition, Honolulu, HI.
Cunningham, C., & Lachapelle, C. (2010). *The impact of Engineering is Elementary (EiE) on students' attitudes toward engineering and science*. Paper presented at the American Society for Engineering Education Annual Conference, Louisville, KY.
Custer, R., & Daugherty, J. (2009). Professional development for teachers of engineering: Research and related activities. *The Bridge, 39*(3), 18–24.
Dayton Regional STEM Center (2011). STEM Education Quality Framework. Available http://daytonregionalstemcenter.org/wp-content/uploads/2012/07/Final-Framework-copyrighted.pdf.
Diamond, K. E., Justice, L. M., Siegler, R. S., & Snyder, P. A. (2013). *Synthesis of IES research on early intervention and early childhood education* (NCSER 2013–3001). Washington, DC: National Center for Special Education Research, Institute of Education Sciences, U.S. Department of Education.
Diamond, K. E., & Powell, D. R. (2011). An iterative approach to the development of a professional development intervention for head start teachers. *Journal of Early Intervention, 33*(1), 75–93.
Dubosarsky, M., Murphy, B., Roehrig, G., Frost, L. C., Jones, J., Carlson, S. P. ... Bement, J. (2011). Animal tracks on the playground, minnows in the sensory table: Incorporating cultural themes to promote preschoolers' critical thinking in American Indian Head Start classrooms. *Young Children, 66*(5), 20–29.
Duncan, F. J., Dowsett, C. J., Claessent, A., Magnuson, K., Huston, A. C., Klebanove, P., et al. (2007). School readiness and later achievement. *Developmental Psychology, 43*(6), 1428–1446.
Eccles, J. S., Jacobs, J. E., & Harold, R. D. (1990). Gender role stereotypes, expectancy effects, and parents' socialization of gender differences. *Journal of social issues*, *46*(2), 183–201.
Eshach, H. (2006). *Science literacy in primary schools and pre-schools*. Dordrecht, The Netherlands: Springer.
Evangelou, D. (2010). Child development perspective in engineering education: Why STEM now? *Early Childhood Research and Practice,* 12(2).
Federal Interagency Forum on Child and Family Statistics. (2013). *America's children: Key national indicators of well-being*. Washington, DC: U.S. Government Printing Office.
Fennema, E. H., & Sherman, J. A. (1977). Sex-related differences in mathematics achievement, spatial visualization, and sociocultural factors. *American Educational Research Journal, 14,* 51–71.
French, L. (2004). Science as the center of a coherent, integrated early childhood curriculum. *Early Childhood Research Quarterly, 19,* 138–149.
Gelman, S. A. (1999). *Concept development in preschool children*. Dialogue on early childhood science, mathematics, and technology education. Washington, DC: Project 2061, American Association for the Advancement of Science.

Genishi, C., & Goodwin, A. L. (2008). *Diversities in early childhood education: Rethinking and doing*. New York: Routledge.
Ginsburg, H., Lee, J., & Boyd, J. (2008). Mathematics education for young children: What it is and how to promote it. *Social Policy Report, 22,* 1–23.
Gorey, K. M. (2001). Early childhood education: A meta-analytic affirmation of the short- and long-term benefits of educational opportunity. *School Psychology Quarterly, 16*(1), 9–30.
Gormley, W. T., Phillips, D., & Gayer, T. (2008). Preschool programs can boost school readiness. *Science, 320,* 1723–1724.
Greenfield, D. B., Jirout, J., Dominguez, X., Greenberg, A., Maier, M., & Fuccillo, J. (2009). Science in the preschool classroom: A programmatic research agenda to improve science readiness. *Early Education and Development, 20*(2), 238–264.
Hill, C., Corbett, C., & St. Rose, A. (2008) *Why so few? Women in Science, Technology, Engineering, and Mathematics*. AAUW http://www.aauw.org/files/2013/02/Why-So-Few-Women-in-Science-Technology-Engineering-and-Mathematics.pdf.
Hjalmarson, M., Diefes-Dux, H. A., & Moore, T. (2008). Designing model development sequences for engineering. In J. Zawojewski, K. Bowman, & H. A. Diefes-Dux (Eds.), *Mathematical modeling in engineering education: Designing experiences for all students*. Roterdam, the Netherlands: Sense Publishers.
Hsu, M. C., Purzer, S., & Cardella, M. E. (2011). Elementary teacher's views about teaching design, engineering, and technology. *Journal of Pre-College Engineering Education Research, 1,* 31–39.
Jacobs, J. E., & Eccles, J. S. (1985). Gender differences in math ability: The impact of media reports on parents. *Educational researcher*, *14*(3), 20–25.
Jenniches, S., & Didion, C. (2009). Engineergirl! A website to interest girls in engineering. *The Bridge, 39*(3), 38–44.
Katehi, L., Pearson, G., & Feder, M. (2009). The status and nature of K-12 engineering education in the United States. *The Bridge, 39*(3), 5–10.
Katz, L. G. (2010). STEM in the early years. *Early Childhood Research and Practice*, 12(2), 11–19. Retrieved from: http://ecrp.uiuc.edu/beyond/seed/katz.html.
Kinzie, M. B., Pianta, R. C., Kilday, C. R., McGuire, P. R., & Pinkham, A. M. (2009, March). *Development of curricula, teacher supports, and assessments for pre-kindergarten mathematics and science*. Paper presented at the annual meeting of the Society for Research on Educational Effectiveness (SREE), Washington, D.C.
Kolodner, J. L., Camp, P. J., Crismond, D., Fasse, B., Gray, J., Holbrook, J., et al. (2003). Problem-based learning meets case-based reasoning in the middle-school science classroom: Putting learning by design into practice. *Journal of the Learning Sciences, 12,* 495.
Levitch, A. & Gable, S. (2016). *Reducing stereotyping in the preschool classroom*. Available http://extension.missouri.edu/hes/childcare/reducestereotype.htm. Accessed February 18, 2016.
Lindeman, K. W., Jabot, M., & Berkley, M. T. (2013). The role of STEM (or STEAM) in the early childhood setting. In L. E. Cohen & S. Waite-Stupiansky (Eds.), *Learning Across the Early Childhood Curriculum,* 17, 95–114. *Advances in Early Education and Day Care series*. Bingley, UK: Emerald Group.
Lynch, S. (2001). "Science for all" is not equal to "One size fits all": Linguistic and cultural diversity and science education reform. *Journal of Research in Science Teaching, 38*(5), 622–627.
Mehalik, M. M., Doppelt, Y., & Schunn, C. D. (2008). Middle-school science through design-based learning versus scripted inquiry: Better overall science concept learning and equity gap reduction. *Journal of Engineering Education, 97,* 71–85.
Moomaw, S. (2012). STEM begins in the early years. *School Science and Mathematics, 112,* 57–58.
National Academy of Engineering. (2006). *Tech tally: Approaches to assessing technological literacy*. Washington, DC: The National Academies Press.
National Academy of Engineering, & National Research Council. (2009). *Engineering in K-12 education: Understanding the status and improving the prospects*. Washington, D.C.: The National Academies Press.

National Research Council. (2009). *Mathematics learning in early childhood: Paths toward excellence and equity*. Washington, DC: National Academy Press.
National Research Council. (2002). *Scientific research in education*. In R. J Shavelson & L Towne (Eds.), Committee on scientific principles for education research. Washington, DC: National Academy Press.
NGSS Lead States. (2013). *Next generation science standards: For states, by states*. Washington, DC: The National Academies Press.
NSB (National Science Board). (2007). *A national action plan for addressing the critical needs of the U.S. science, technology, education, and mathematics education system*. Washington, D.C.: National Science Foundation.
New, R. S. (1999). Playing fair and square: Issues of equity in preschool mathematics, science, and technology. *Dialogue on Early Childhood Science, Mathematics, and Technology Education: Fostering High Quality Programs*. Washington, DC: Project 2061, American Association for the Advancement of Science. http://www.project2061.org/publications/earlychild/online/fostering/new.htm.
Novak, J. D. (1977). *A theory of education*. Ithaca, NY: Cornell University Press.
O'Donnell, C. (2008). Defining, conceptualizing, and measuring fidelity of implementation and its relationship to outcomes in K-12 curriculum intervention research. *Review of Educational Research, 78,* 33–84.
Piaget, J. (1961). *The child's conception of number*. New York: Norton.
Powell, D. R., Son, S., File, N., & San Juan, R. R. (2010). Parent-school relationships and children's academic and social outcomes in public school pre-kindergarten. *Journal of School Psychology, 48*(4), 269–292.
Roberts, L. C., & Hill, H. T. (2003). Using children's literature to debunk gender stereotypes. In Carol Copple (Ed.), *A world of difference: Readings on teaching young children in a diverse society* (pp. 125–127).
Samarapungavan, A., Mantzicopoulos, P., & Patrick, H. (2008). Learning science through inquiry in kindergarten. *Science Education, 92*(5), 868–908.
Schunn, C. D. (2009). *How kids learn engineering: The cognitive science perspective*. The Bridge: Linking engineering and society. US: National Academy of engineering of the National Academies.
Seymour, E., & Hewitt, N. M. (1997). *Talking about leaving: Why undergraduates leave the sciences*. Boulder, CO: Westview Press.
Sleeter, C. E., & Grant, C. A. (1999). *Making choices for multicultural education: Five approaches to race, class and gender* (3rd ed.). New Jersey: Merrill.
Sneider, C. I. (Ed.). (2014). *The go-to guide for engineering curricula, preK-5: Choosing and using the best instructional materials for your students*. Corwin Press.
Stoll, J., Hamilton, A., Oxley, E., Eastman, A. M., & Brent, R. (2012). Young thinkers in motion: Problem solving and physics in preschool. *Young Children, 67,* 20–26.
Center, Southern Poverty Law. (1997). Gender learning in early childhood. *Starting small: Teaching tolerance in preschool and the early grades* (pp. 50–54). Author. Reprinted: Montgomery, AL.
Sullivan, A., Kazakoff, E. R., & Bers, M. U. (2013). The wheels on the bot go round and round: Robotics curriculum in pre-kindergarten. *Journal of Information Technology Education: Innovations in Practice, 12,* 203–219.
Swim, J. K. (1994). Perceived versus meta-analytic effect sizes: An assessment of the accuracy of gender stereotypes. *Journal of Personality and Social Psychology, 66,* 21–36.
Torres-Crespo, M. N., Kraatz, E., & Pallansch, L. (2014). From fearing STEM to Playing with It: The natural integration of STEM into the preschool classroom. *Journal of the Southeastern Regional Association of Teacher Educators, 23*(2), 8–16.
Tunks, K. W. (2009). Block play: Practical suggestions for common dilemmas. *Dimensions of Early Childhood, 37,* 3–8.

U.S. Department of Education, National Center for Education Statistics (2007). *The Nation's Report Card: America's high school graduates: Results from the 2005 NAEP High School Transcript Study*, by C. Shettle et al. (NCES 2007-467) Washington, DC: Government Printing Office.
Van Meeteren, B., & Zan, B. (2010). Revealing the work of young engineers in early childhood education. *Early Childhood Research and Practice*. Collected Papers from the SEED (STEM in Early Education and Development) Conference.
Wang, J., Werner-Avidon, M., Newton, L., Randol, S., Smith, B., & Walker, G. (2013). Ingenuity in action: Connecting tinkering to engineering design processes. *Journal of Pre-College Engineering Education Research, 3,* 1–23.
Warren, B., Ballenger, C., Ogonowski, M., Rosebery, A. S., & Hudicourt-Barnes, J. (2001). Rethinking diversity in learning science: The logic of everyday sense making. *Journal of Research in Science Teaching, 38,* 529–552.
Wolfson, W. (n.d.). *Engineering Lens*. Retrieved from: http://www.integratingengineering.org/index.html.
Worth, K. (2010). *Science in early childhood classrooms: Content and process*. Collected Papers from the SEED (STEM in Early Education and Development) Conference. Retrieved from: http://ecrp.uiuc.edu/beyond/seed/worth.html.
Wynn, T., & Harris, J. (2013). Toward a STEM and arts curriculum: Creating the teacher team. *Art Education, 65,* 42–47.
Xie, Y., & Shauman, K. A. (2003). *Women in science: Career processes and outcomes*. Cambridge, MA: Harvard University Press.

Part III
Concluding Points

Chapter 13
Engineering Education in Early Childhood: Reflections and Future Directions

Lyn English

Abstract As the chapters in this book illustrate, engineering is a natural feature of early childhood learning and development. Young children have innate tendencies to participate in engineering activities, displaying sophisticated design and thinking processes in doing so. Teachers need support, however, in facilitating this early development. The need for more developmentally and culturally appropriate curriculum resources remains a pressing concern when trying to implement engineering in a range of early childhood classrooms. Of the numerous areas requiring attention, this chapter touches on just a few in both reflecting on the current scene and suggesting recommendations for advancing the field. These include (a) the need to incorporate both engineering design processes and habits of mind in promoting early engineering learning, (b) the creation of developmentally appropriate experiences that provide pedagogical affordances, and (c) the integration of engineering in STEM education, with extensions to STEAM (science, technology, engineering, the arts, mathematics).

Engineering is a key component of STEM education (science, technology, engineering, mathematics) and yet remains largely neglected in the early and elementary years (Aguirre-Munoz & Pantoya, 2016; English, 2016; Watkins, Spencer, & Hammer, 2014). Given the need to provide some foundation for rectifying the current state of play, this book was developed with contributions from leaders in the field. As the chapters indicate, engineering is a natural feature of early childhood learning and development.

Given the innate tendencies of young children to engage in engineering thinking, teachers need support in facilitating this early development. It seems somewhat paradoxical that beginning engineering is a natural feature of young children's learning, yet the domain has been largely ignored in these informative years. This situation is perhaps not surprising, given research indicating teachers' lack of preparedness,

---

The original version of this chapter was revised: XX entries are removed. The correction to this chapter is available at https://doi.org/10.1007/978-981-10-8621-2_14

L. English (✉)
Queensland University of Technology, Brisbane , Australia
e-mail: l.english@qut.edu.au

L. English and T. Moore (eds.), *Early Engineering Learning*, Early Mathematics Learning and Development, https://doi.org/10.1007/978-981-10-8621-2_13

and hence confidence, in implementing engineering within their classroom curriculum (Crismond & Adams, 2012; Nadelson & Seifert, 2017). Insufficient professional development opportunities and the lack of suitable curriculum resources exacerbate the situation. Although the present chapters provide valuable insights into existing programs and the rich learning experiences they provide, the need for more developmentally and culturally appropriate curriculum resources remains a pressing concern (Early Childhood STEM Working Group, 2017).

Of the numerous foundational engineering components addressed by the authors, I touch on just a few in both reflecting on the current scene and suggesting recommendations for advancing the field. Given the scant literature on the topic (and hence, an impetus for this book), any recommendations are tentative and require further research and application within diverse educational environments. I give consideration to: (a) the need to include both engineering design processes (e.g., Bagiati & Evangelou, Chap. 6; English, King, & Smeed, 2017; Estapa & Tank, 2017; Watkins et al., 2014) and engineering habits of mind in promoting early engineering learning (e.g., Lippard, Riley, & Lamm, Chap. 3; Lucas, Hanson, & Claxton, 2014); (b) the creation of developmentally appropriate experiences that provide pedagogical affordances (Gadanidis, Hughes, Minniti, & White, 2016); and (c) the integration of engineering in STEM education, with extensions to STEAM (science, technology, engineering, the arts, mathematics).

13.1 A Focus on Engineering Design and Habits of Mind

Design processes form a key component of beginning engineering learning, yet engineering design is underrepresented in the early and elementary school years. This is in spite of engineering design being viewed internationally as a foundational, linking process across the STEM disciplines and not just confined to engineering (Lucas, Claxton, & Hanson, 2014; Claxton & Lucas, 2015; *Next Generation Science Standards* [NGSS], 2014). Various examples and representations of engineering design appear in the chapters in this book, with a general focus on iterative thinking, that is, developing a problem solution, testing it, learning from what does not go well, and trying again. Working in teams and communicating current ways of thinking are also highlighted (Bryan, Moore, Johnson, & Roehrig, 2016; English & King, 2015; King & English, 2017).

Engineering "habits of mind" or engineering thinking underlines design processes and includes systems thinking, innovative problem finding and solving, visualizing, and collaborating and communicating (Lucas et al., 2014; English & Gainsburg, 2016). There are several variations of these engineering habits of mind including Katehi, Pearson, and Feder's (2009, p. 7) focus on systems thinking, creativity, optimism, collaboration, communication, and ethical considerations. Systems thinking and creative problem solving appear to be frequently cited as foundational not only in engineering learning but also across the spectrum of STEM education (e.g., English & King, 2016). The importance of developing engineering habits of mind is emphasized in several chapters including that of Lippard, Riley, and Lamm, who point out that such thinking processes are not a set curriculum. Rather, these processes should be viewed as "developmental outcomes that arise from children's meaningful interactions with engineering concepts and activities". These habits of mind should thus be embedded and fostered within existing classroom curricula and instructional practices. The present chapters provide numerous examples of how this approach could be adopted, with many such approaches capitalizing on children's literature, to which I return.

By incorporating engineering design and engineering habits of mind within the regular classroom curriculum, young children's conceptual understanding across the STEM disciplines can be fostered. Ultimately, such conceptual learning should lead to the development of what McKenna (2014) refers to as "adaptive expertise" (p. 232) involving applying what one learns to new situations. In other words, engineering-based problems have the potential to encourage young students to "learn from and about the problem, while continually reflecting on, and possibly reshaping, prior knowledge and experiences" (McKenna, 2014, p. 232). Early engineering-based problems that embed design constraints and draw on meaningful interdisciplinary contexts can assist learners to recognize what knowledge they need to apply to a new situation. For example, Tank et al. (Chap. 9) illustrate how their *PictureSTEM* curricula introduce young students to STEM-based problems where they initially define and develop a better understanding of the problem being addressed, including the scientific and mathematical knowledge embedded in the problem. Such knowledge is further developed as students engage in associated mathematics and science activities, and subsequently learn what is needed in applying design processes to problem solution. Furthermore, students develop knowledge of ways in which their design can provide an acceptable and optimum solution.

13.2 Developmentally Appropriate Experiences with Pedagogical Affordances

Implementing early engineering experiences that capitalize on both the learning potential of young children and the appealing, multidisciplinary contexts that support engineering can be challenging but need not be. As repeatedly indicated, engineering links the STEM disciplines and is an ideal vehicle for advancing the natural curiosity and problem-solving skills that characterize early childhood (e.g., Elkins, Sullivan, & Bers, Chap. 11). Knowing how to structure STEM activities that lift the profile of engineering, that are geared toward the child's world, and that cater for different developmental levels is paramount. The present chapters provide examples of guidelines that can be used to structure such experiences, including the "design parameters" of Cunningham and Lachapelle (Chap. 8) and the activity guidelines proposed by Dubosarsky et al. (Chap. 12).

Studies conducted by Gadanidis et al. (2016) have revealed how young students' mathematical learning can be enriched and extended through designing activities that have core features of: (a) low floors, (b) high ceilings, (c) wide walls, and (d)

Table 13.1 Learning affordances for early engineering experiences

Low floor	Enables young learners to engage in solving an engineering-based problem for which they have minimal conceptual knowledge and can tackle the problem at their entry or readiness level
High ceilings	Afford learners opportunities to extend their thinking and learning, often to work with ideas that are beyond their year or grade level. The focus of the activity then becomes one of learning or idea generation
Wide walls	Encourages students to share and communicate their learning not only within the classroom but also beyond, as they convey to others the conceptual surprises they have experienced
Conceptual surprise	Develops excitement and surprise as new ideas are uncovered

generation of conceptual surprises for both the learners and their teachers. These pedagogical affordances apply equally to integrated STEM experiences that develop foundational engineering learning. As applied to such experiences, these affordances are characterized in Table 13.1.

Engineering experiences that display these learning affordances are ideal for young children, who are readily capable of dealing with challenging activities that extend their thinking (e.g., English, 2013, 2016; Ginsburg, 2016; Ginsburg, Cannon, Eisenband, & Pappas, 2006). As the engineering education literature has shown, when rich and meaningful contexts are provided, young learners can engage in engineering activities and display impressive learning (Portsmore & Milto, Chap. 10; Portsmore, Watkins, & McCormick, 2012; Watkins et al., 2014). The second principle of Portsmore and Milto's approach draws on this research, with their programs utilizing meaningful contexts that both capitalize on young learners' abilities and enable them to develop their problem-solving skills. Such contexts, illustrated in Portsmore and Milto' chapter, enable young learners to engage in engineering activities with minimal instruction (Portsmore et al., 2012; Watkins et al., 2014). The teacher serves as a facilitator, responding appropriately to students' discussions and solution efforts. An important aspect of this facilitation is the structure of the activity. By incorporating the foregoing learning affordances into activity design, students' learning potential can be targeted with all students nurtured to engage meaningfully and achieve a sense of purpose and accomplishment.

13.3 Integrating Early Engineering Learning Within a STEAM Curriculum

With the increased attention to improving STEM education, and more recently, extending this learning to STEAM (incorporating the arts), new avenues have opened to integrating engineering experiences. As the present chapters have shown, engineering fits in naturally with other disciplines in early childhood programs, despite the challenges in developing integrative curricula (e.g., English et al., 2017; Johnson,

Peters-Burton, & Moore, 2016; Nadelson & Seifert, 2017). Compared to other educational areas, however, research remains limited especially with respect to the positive learning outcomes that result from STEM integration (Pearson, 2017). Although not without its difficulties, integrating engineering within early STEM programs would appear to present fewer challenges than in later grade levels where such approaches can become more complex and inflexible due to timetable and other restrictions.

It is not the intent of this section to review the numerous articles that have appeared on STEM integration in recent years (e.g., English et al., 2017; Honey, Pearson, & Schweingrubwe, 2014; Johnson et al., 2016). Rather, I refer to Nadelson and Seifert's (2017) article in which they make the important point that our "age of synthesis targets new domains of expertise." Such expertise includes applying seemingly unrelated information from different disciplines, dealing with the rapid emergence of new knowledge and technologies, and preparing for transdisciplinary careers. Nadelson and Seifert's (2017) definition of integrated STEM appears appropriate for addressing the inclusion of engineering within early learning programs:

> We define integrated STEM as the seamless amalgamation of content and concepts from multiple STEM disciplines. The integration takes place in ways such that knowledge and process of the specific STEM disciplines are considered simultaneously without regard to the discipline, but rather in the context of a problem, project, or task (p. 221).

Such a perspective does not lose sight of the importance of foundational discipline-specific knowledge, however. STEM learning requires a mix of integrated and singular disciplinary experiences; finding the right balance is a key consideration in curriculum development (e.g., English et al., 2017; Johnson et al., 2016). In the early childhood years, integrated learning occurs naturally as children undertake class projects as well as solve problems beyond the classroom. For example, in designing a hutch for their rabbit, children would need to consider the shape and dimensions of their structure, suitable materials to use, ways in which their rabbit would enter and exit the hutch, ways of insulating the hutch, appropriate food supplies, how to arrange these within the hutch, and so on. Simply providing such experiences without considering how the respective STEM components come together is insufficient. In other words, integrated STEM needs to be crafted, with special attention given to how engineering can be highlighted. The use of an appealing context, frequently supported through the literature, is an essential feature of crafting early engineering activities.

Also contributing to such a context is the aesthetics domain, which is an under-represented yet significant component of engineering and engineering education. Indeed, the recent expansion of STEM to incorporate the arts resulting in the STEAM acronym (English et al., 2017; Sinclair, 2006) has implications for young children's engineering experiences.

Research on aesthetic approaches to teaching children mathematics (e.g., Sinclair, 2006) has revealed how young learners can find beauty in the mathematics they experience, rather than being preoccupied with issues pertaining to "correctness" and to "passing tests." Fortunately, early engineering provides opportunities for aesthetically pleasing experiences where young learners can investigate motivating real-world problems and create artifacts that hold beauty and personal meaningfulness for them. As Sinclair (2006) aptly stated, "The aesthetics judgements they [children] make contrast to the judgments they may make about whether or not their work is procedurally correct or acceptable to the teacher" (p. 7). By its very nature, early engineering experiences are aesthetically enticing; indeed, we can find endless examples of engineered creations in our world that are designed to be both functional and aesthetically appealing. The Sydney Opera House (Australia) is a case in point, where the architect Jorn Utzon's masterpiece was inspired by nature, its forms, functions, and colors. His design for the Opera House was influenced by bird wings and the shape and form of clouds, shells, palm trees, and walnuts. In drawing on nature, Utzon designed a structure that was functional, sustainable, efficient, and beautiful.

Hand-in-hand with aesthetics is children's literature. Children's picture storybooks naturally utilize the aesthetics domain and serve as powerful contexts for early engineering learning. There appear two fundamental ways in which the literature, specifically children's picture storybooks, can be used in implementing engineering activities: general storybooks that do not specifically target engineering but can serve as a rich springboard for engineering experiences, and engineering-centered storybooks that directly target the engineering domain, together with the STEM disciplines more broadly.

13.4 General Children's Literature: Non-engineering Specific

Rich opportunities for early STEM learning abound in children's literature (e.g., Columbia, Kim, & Moe, 2005; Luedtke & Sorvaag, 2017). The chapters in this book provide many comprehensive examples of how the literature in general can provide stimulating scenarios that are open to exploring engineering ideas. Portsmore and Milto (Chap. 10) make the important point with respect to their *Novel Engineering* program, namely a myriad of books can stimulate engineering design challenges. Children are prompted to draw on the information in the story lines to identify engineering problems, such as imagining that the characters need their assistance in solving a particular dilemma. Children can identify the nature of the dilemma, the constraints that might exist in finding possible solutions (e.g., limited materials or tools available, time constraints, ways to avoid predators), and approaches to tackling the problematic situation. As Portsmore and Milto highlight, by leveraging existing storybooks teachers can easily integrate *Novel Engineering* within their existing curricula and thus customize their approach, especially when they might lack the confidence or experience in implementing this new domain. Furthermore, the integration of early engineering experiences within existing curricula can overcome concerns regarding time constraints that might occur in an already crowded school day.

Tank et al. (Chap. 9) provide numerous other examples of how the literature can set the scene for a range of engineering experiences within their *PictureSTEM* program. Furthermore, these experiences can be readily combined with teaching phonemes and word recognition. The use of an engineering design process provides an anchor and an important link for each lesson within a *PictureSTEM* unit, thus helping children see how their different learning areas are connected and contribute collectively to the solution of real-world problems. By broadening the ways in which children's literature is used, increased opportunities arise for young children to explore engineering in their world. These opportunities can be enriched and supplemented with children's books that specifically target engineering and the STEM domains more broadly.

13.5 Engineering-Centered Children's Literature

Early childhood literature focusing primarily on engineering experiences is not prevalent in the field and is a comparatively recent development. Yet research has indicated that combining the engineering-centered literature with enhanced classroom conversations and discussion can be effective tools for broadening children's participation in engineering education (Aguirre-Munoz & Pantoya, 2016; Cunningham, Lachapelle, & Davis, Chap. 4). The *Engineering is Elementary* program, developed by Lachapelle and Cunningham (e.g., 2014) at the Boston Museum of Science, is a well-known case in point. Their program incorporates storybooks that provide examples of characters solving specific engineering problems (as illustrated in Chap. 4). The first principle listed in their set of design principles pertains to a *narrative context*, where the reason for the design challenge must make sense and be meaningful within children's world; reading and communication are foundational to these elementary engineering experiences.

Other engineering-centered children's literature includes the series developed by King and Johnston (e.g., 2012, 2014), which takes children on a journey with a mix of human and animal characters as they solve appealing problems. Children are encouraged to contribute to the solution process as each new scenario is presented and hypotheses considered. Problem scenarios are primarily engineering-based but also incorporate the other STEM areas. The series is aesthetically pleasing as the artwork is not only designed to captivate young children's interest but is also a key component of the problems presented and possible solutions entertained. Figure 13.1 provides an example of the way in which the artwork serves these two primary purposes.

13.6 Future Actions

Introducing engineering education in the early childhood years has yet to gain substantial momentum, at least across many countries. The chapters in this book highlight

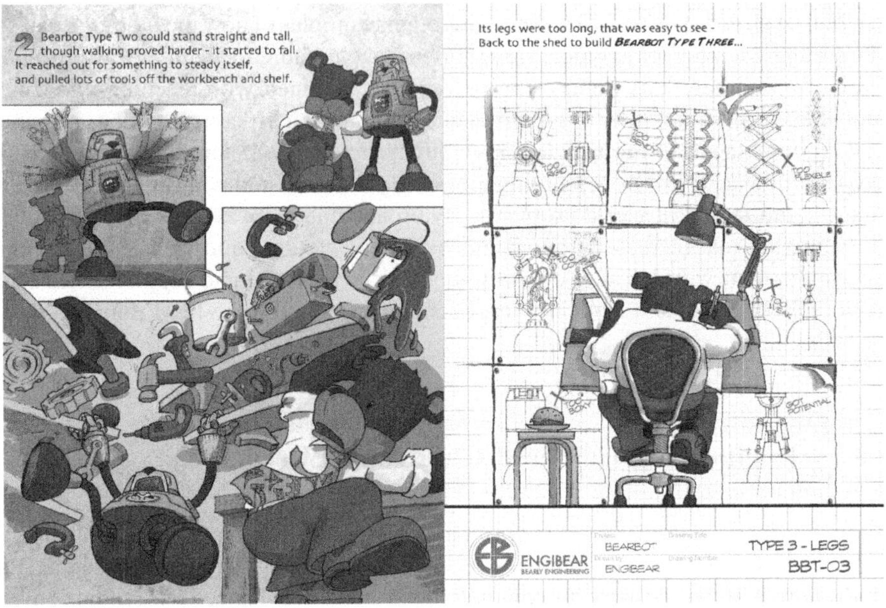

Fig. 13.1 An illustration from *Engibear's Dream* (with kind permission from King & Johnston, 2012)

possible reasons for this situation and offer numerous recommendations for advancing the field more broadly. Although far from exhaustive, two key areas that appear in need of attention with respect to research, policy, and curriculum development include: increasing awareness of young children's competencies in early engineering and enhancing teacher resources and professional development opportunities.

13.7 Increasing Awareness of Young Children's Engineering Capabilities

Despite the almost commonplace use of the STEM acronym, the E often remains silent, especially in the beginning and elementary years. One could even question cynically whether its role is to complete the acronym rather than signal a core learning area. The scarce appearance of engineering within early childhood curricula is compounded by beliefs that young children are incapable of tackling the domain. Such a situation is perhaps not surprising given that young children's learning potential still does not receive the recognition warranted (English, 2016; Ginsburg, 2016). Their capabilities are often masked by standardized tests that explore a narrow range of reasoning processes and problem-solving skills. As Ginsburg and his colleagues have argued over many years (e.g., Ginsburg et al., 2006), a good deal of research on early mathematics learning has been restricted to an analysis of children's actual developmental level; this has failed to illuminate their potential for learning under stimulating conditions that challenge their thinking: Research that focuses on children's current knowledge and reasoning is inadequate. Other researchers have likewise warned of how common misconceptions and underestimations of young learners' developmental capabilities can hamper the scope of the learning experiences we provide (Moss, Bruce, & Bobis, 2016; Perry & Dockett, 2008). Although these cases refer specifically to mathematics learning, they are equally applicable to the engineering domain, probably more so, given that engineering education is typically viewed as a tertiary level or secondary school area of study (DiFrancesca, Lee, & McIntyre, 2014).

Unfortunately, there exist few established measures that can reveal the engineering capabilities of young learners. Purzer and Douglas (Chap. 7) offer a detailed assessment model for gaining insights into emerging engineering thinking and design skills, with their "Mosaic Framework" comprising three key principles: (a) a base of essential learning goals and objectives aligned with the classroom curriculum, (b) embedded assessment tools that effectively capture evidence of learning, and (c) professional development that incorporates the interpretation and integration of multiple assessment forms.

One such important form of assessment is guided observation of children's experimentations with blocks and other common hands-on equipment. A wealth of insights into the foundations of early engineering skills can be gained from observing young children at play. The problem remains, however, of what to look for in such observations especially for educators who are new to the domain. Examples of observation protocols are presented in this book, such as that offered by Bagiati and Evangelou (Chap. 6) in an observational study of children's informal play. As the authors illustrate in their detailed protocol, children's play provides a valuable setting for the learning and development of engineering thinking. The authors present an important caveat, however, with respect to authenticity in both early childhood curricula and engineering content: "Any attempt to bring engineering into early education must be accompanied by sincere and systematic efforts to educate teachers so that they may take on the responsibility of introducing engineering in their classrooms effectively and with confidence". Unfortunately, the professional development experiences required are scarce, which impede the implementation of beginning engineering experiences.

13.8 Enhancing Teacher Resources and Professional Development Opportunities

Educating teachers in early engineering learning is made all the more difficult by the limited available resources, as Bagiatti and Evangelou (Chap. 6) noted. Without an increase in the number of worthwhile resources, the advancement of early engineering will be slowed. Several examples of specifically designed programs for younger learners are presented in the chapters, with a focus on developmentally appropriate experiences that integrate effectively with existing curricula (e.g., Cunningham & Lachapelle, Chap. 8).

As argued earlier in this concluding chapter, early engineering resources need to feature pedagogical affordances that not only enable all children to participate, but also capitalize on their learning potential. The curriculum design parameters that guide the development of engineering experiences in Cunningham and Lachapelle's chapter, for example, illustrate the opportunities that are afforded. Children can approach their problems in a variety of ways and produce valid and creative solutions. In so doing, they can improve constructively on failure or dissatisfaction, reflect on their progress, and generate and test new ideas, approaches, and solutions.

As previously indicated, the *PictureSTEM* program (Chap. 9) is another example of resources that teachers can access. The program provides a detailed set of units each comprising paired reading and STEM lessons. The literature is integrated within STEM and provides important context, background knowledge, and linking concepts. The engineering design context is a fundamental linking component across the units. With the detailed content background, rationale, and lesson structures, the program can also serve as a template or guideline for the development of other engineering-based STEM programs. As with each of the programs featured in this book, high-quality professional development is an integral component.

There is more to professional development, however, than simply providing one-off sessions to introduce a particular program. Estapa and Tank (2017) report research that highlights the need for sustained, coherent, and collaborative teacher programs that develop more in-depth understandings of the STEM content domains and various ways of integrating engineering within the disciplines. One of the problems with many professional development programs is the uncertainty of what specific content and skills are most effective and how these can be conveyed to upscale the adoption of integrated STEM programs (O'Brien, Karsnit, Sandt, Bottomley, & Parry, 2014).

An interesting approach adopted in Estapa and Tank's study engaged triads comprising a classroom teacher, preservice teacher, and an engineering fellow in professional development experiences centered on STEM concepts and the use of engineering design. Ways in which this learning was subsequently implemented in the classroom were investigated. One of the findings was the difficulty experienced in forging STEM connections in practice, with the disciplines tending to remain siloed. Furthermore, engineering content was perceived as a particular skill or practice rather than as academic content. Assisting teachers in better understanding the nature and role of engineering learning, together with effective planning and the enactment of integrated STEM lessons, appears a key area for future research and action.

The foregoing recommended areas for action are by no means exhaustive. Nevertheless, they do highlight some of the key issues demanding attention, not only from researchers and early childhood educators, but also from policy and curriculum planning personnel. This book is one attempt to provide research-based, practical examples of how early childhood programs can capture and build on young children's natural abilities and interest in the field. It is hoped that the chapters provide the stimulus for further action in such an important domain.

References

Aguirre-Munoz, Z., & Pantoya, M. L. (2016). Engineering literacy and engagement in kindergarten classrooms. *Journal of Engineering Education, 105*(4), 630–654. https://doi.org/10.1002/jee.20151.

Bryan, L. A., Moore, T. J., Johnson, C. C., & Roehrig, G. H. (2016). Integrated STEM education. In C. C. Johnson, E. E. Peters-Burton, E. E., & T. J. Moore (Eds.), *STEM road map: A framework for integrated STEM education*. New York, NY, Routledge.

Claxton, G., & Lucas, B. (2015). *Educating Ruby: What out kids really need to learn*. Bethel, CT: Crown House Publishing Limited.

Crismond, D. P., & Adams, R. S. (2012). The informed design teaching and learning matrix. *Journal of Engineering Education, 101*(4), 738–797.

Columbia, L., Kim, C., & Moe, A. (2005). *The power of picture books in teaching math and science: Grades preK-8*. Scottsdale, AZ: Holcomb Hathaway Publishers Inc.

DiFrancesca, D., Lee, C., & McIntyre, E. (2014). Where is the "E" in STEM for young children? *Issues in Teacher Education*, Spring, 49–64.

Early Childhood STEM Working Group (2017). *Early STEM matters: Providing high-quality STEM experiences for all young learners*. http://ecstem.uchicago.edu.

English, L. D. (2013). Reconceptualising statistical learning in the early years. In L. D. English & J. Mulligan (Eds.), *Reconceptualising early mathematics learning* (pp.67–82). Dordrecht: Springer.

English, L. D. (2016). STEM education K-12: Perspectives on integration. *International Journal of STEM Education, 3*(1). http://dx.doi.org/10.1186/s40594-016-0036-1.

English, L. D., & Gainsburg, J. (2016). Problem solving in a 21st—century mathematics curriculum. In L. D. English & D. Kirshner (Eds.), *Handbook of international research in mathematics education* (3rd ed., pp. 313–335). New York: Routledge.

English, L. D., & King, D. T. (2015). STEM learning through engineering design: Fourth-grade students' investigations in aerospace. *International Journal of STEM Education, 2* (14). https://doi.org/10.1186/s40594-015-0027-7. URL: http://www.stemeducationjournal.com/content/2/1/14.

English, L. D., & King, D. T. (2016). Designing earthquake resistant buildings. iSTEM section, *Teaching Children Mathematics, 23*(1), 47–50.

English, L. D., King, D., & Smeed, J. (2017). Advancing integrated STEM learning through engineering design: Sixth-grade students' design and construction of earthquake resistant buildings. *The Journal of Educational Research, 110*(3), 255–271.

Estapa, A. T., & Tank, K. M. (2017). Supporting integrated STEM in the elementary classroom: A professional development approach centered on an engineering design challenge. *International Journal of STEM Education, 4*(6). https://doi.org/10.1186/s40594-017-0058-3.

Gadanidis, G., Hughes, J. M., Minniti, L., & White, B. J. G. (2016). Computational thinking, grade 1 students and the binomial theorem. *Digital Experiences in Mathematics Education*. https://doi.org/10.1007/s40751-016-0019-3.

Ginsburg, H. P. (2016). Helping early childhood educators to understand and assess young children's mathematical minds. *ZDM Mathematics Education, 48*(7).

Ginsburg, H. P., Cannon, J., Eisenband, J. G., & Pappas, S. (2006). Mathematical thinking and learning. In K. McCartney & D. Phillips (Eds.), *Handbook of early child development* (pp. 208–230). Oxford, England: Blackwell.

Honey, M., Pearson, G., & Schweingruber, H. (Eds.). (2014). *STEM integration in K-12 education: Status, prospects, and an agenda for research*. Washington, DC: National Academies Press.

Johnson, C. C., Peters-Burton, E. E., & Moore, T. J. (2016). *STEM road map: A framework for integrated STEM education*. New York: NY, Routledge.

Katehi, L., Pearson, G., & Feder, M. (2009). K-12 engineering education in the United States. In *The Bridge: Linking engineering and society*. Washington: National Academy of Engineering.

King, D. T., & English, L. D. (2017). Engineering design in the primary school: Applying STEM concepts to build an optical instrument. *International Journal of Science Education, 18,* 2762–2794. https://doi.org/10.1080/09500693.2016.1262567.

King, A., & Johnston, B. (illustrator). (2012). *Engibear's dream*. Frenchs Forest: Little Steps Publishing.
King, A., & Johnston, B. (illustrator). (2014). *Engibear's bridge*. Frenchs Forest: Little Steps Publishing.
Lachapelle, C. P., & Cunningham, C. M. (2014). Engineering in elementary schools. In S. Purzer, J. Strobel, & M. Cardella (Eds.), *Engineering in pre-college settings: Synthesizing research, policy, and practices* (pp. 61–88). Lafayette, IN: Purdue University Press.
Lucas, B., Hanson, J., & Claxton, G. (2014). *Thinking like an engineer: Implications for the education system. Royal Academy of Engineers*. www.raeng.org.uk/thinkinglikeanengineer.
Luedtke, M., & Sorvaag, K. (2017). Using children's literature to enhance math instruction in K-8 classrooms. In L. Jao & N. Radakovic (Eds.), *Transdisciplinarity in mathematics education*. Springer.
McKenna, A. F. (2014). Adaptive expertise and knowledge fluency in design and innovation. In A. Johri, A., & B. M. Olds (Eds.), *Cambridge handbook of engineering education research* (pp. 227– 242). New York: Cambridge University Press.
Moss, J., Bruce, C. D., & Bobis, J. (2016). Young children's access to powerful mathematics ideas: A review of current challenges and new developments in the early years. In L. D. English & Kirshner, D. (Eds.). *Handbook of international research in mathematics education,* (3rd ed., pp. 153–190). New York: Routledge.
Nadelson, L. S., & Seifert, A. L. (2017). Integrated STEM defined: Contexts, challenges, and the future. *The Journal of Educational Research, 110*(3), 221–223.
O'Brien, S., Karsnitz, J., Sandt, S., Bottomley, L., & Parry, E. (2014). Engineering in preservice teacher education. In S. Purzer, J. Strobel, & M. Cardella (Eds.), *Engineering in pre-college settings: Synthesizing research, policy, and practices* (pp. 277–300). IN: Purdue University Press.
Pearson, G. (2017). National academies piece on integrated STEM. *The Journal of Educational Research, 110*(3), 224–226.
Perry, B., & Dockett, S. (2008). Young children's access to powerful mathematical ideas. In L. D. English (Ed.), *Handbook of international research in mathematics education* (2nd ed., pp. 75–108). NY: Routledge.
Portsmore, M., Watkins, J., & McCormick, M. (2012). Planning, drawing and elementary students in an integrated engineering design and literacy activity. Paper presented at the 2nd *P-12 Engineering and Design Education Research Summit*. Washington, DC.
Sinclair, N. (2006). The aesthetic sensibilities of mathematicians. In N. Sinclair, D. Pimm, & W. Higggins (Eds.), *Mathematics and the aesthetic: New approaches to an ancient affinity* (pp. 87–104). New York, NY: Springer.
Watkins, J., Spencer, K., & Hammer, D. (2014). Examining young students' problem scoping in engineering design. *Journal of Pre-College Engineering Education Research, 4*(1), 43–53. https://doi.org/10.7771/10.7771/2157-9288.1082.

Correction to: Early Engineering Learning

Lyn English and Tamara Moore

Correction to: L. English and T. Moore (eds.), *Early Engineering Learning*, Early Mathematics Learning and Development, https://doi.org/10.1007/978-981-10-8621-2

In the original version of the book, the following belated corrections have been incorporated: XX entries in chapters 1 and 13 has been removed in both the chapters. The erratum chapters and book have been updated with the changes.

The updated versions of these chapters can be found at
https://doi.org/10.1007/978-981-10-8621-2_1
https://doi.org/10.1007/978-981-10-8621-2_13

L. English and T. Moore (eds.), *Early Engineering Learning*, Early Mathematics Learning and Development, https://doi.org/10.1007/978-981-10-8621-2_14

Index

L. English and T. Moore (eds.), *Early Engineering Learning*, Early Mathematics Learning and Development, https://doi.org/10.1007/978-981-10-8621-2

The manufacturer's authorised representative in the EU is Springer Nature Customer Service Centre GmbH, Europaplatz 3, 69115 Heidelberg, Germany. If you have any concerns regarding our products, please contact ProductSafety@springernature.com

Printed and bound by CPI Group (UK) Ltd, Croydon, CR0 4YY
30/09/2026
02225447-0006